ENVIRONMENTAL MANAGEMENT SY

Edited by Paul Sharratt

INSTITUTION OF CHEMICAL ENGINEERS

Published by
Institution of Chemical Engineers,
Davis Building,
165–189 Railway Terrace,
Rugby, Warwickshire CV21 3HQ, UK

Copyright © 1995 Institution of Chemical Engineers
A Registered Charity

ISBN 0 85295 363 1

The symposium upon which this book is based was organized by the Institution
of Chemical Engineers North Western Branch. The event was co-sponsored by
the Society of Chemical Industry Environment & Water Group and the
Institution of Mechanical Engineers Environmental Engineering Group, and the
Institution of Chemical Engineers Water: Preparation, Use and Cleanup Subject
Group, Environmental Protection Subject Group and Safety, Health and
Environment Department.

I MECH E

Cover photograph: pupils from Rawthorpe Junior School in the wild flower meadow at
Zeneca Huddersfield Works. (Courtesy of *The Huddersfield Examiner*.)

Printed in the United Kingdom by Galliard (Printers) Ltd, Great Yarmouth.

PREFACE

Management of the environmental performance of companies in the process industries has become a pressing problem in recent years. That pressure has come from the public, changes in legislation, rising costs for waste disposal and worries about liability (both within companies and financial bodies). This has resulted in rapid developments in the systems within organizations which co-ordinate and control activities with environmental impact. It is these 'environmental management systems' which are the subject of this book. It brings together a range of approaches and experience relevant to environmental management systems (EMS) for the process industries. It is not a complete treatment — and could not be in this complex and rapidly developing field. The objective is to provide practical guidance through the experiences of environmental professionals working on real problems.

The chapters are grouped by topic. The first three chapters discuss some of the issues which must be addressed by EMS, including regulatory and public concerns. They also touch on some of the problems which can occur if systems are inadequate. The next two chapters provide outlines of the requirements of BS7750 — especially with respect to the accreditation process. Forthcoming developments in international standards are also discussed. Case studies, which form the third topic, describe the approaches taken by various companies to the development of EMS. Finally, tools which address some of the tasks implicit in environmental management are presented. These tasks include risk assessment and management, data handling, methods for assessment of environmental impact and monitoring of releases.

There are several recurrent themes, reflecting some of the current problem areas in EMS. Perhaps the most serious of these is the lack of any cost-effective means of defining environmental impact which is acceptable to everybody. This makes the demonstration of 'continual improvement', as required by some published or proposed standards, rather difficult. It also causes problems with concepts such as best practicable environmental option (BPEO) and best available techniques not entailing excessive cost (BATNEEC). The views of industry and the Her Majesty's Inspectorate of Pollution (HMIP) appear to differ widely in this area. Indeed, two chapters provide on the one hand HMIP's latest contribution to the debate (Chapter 16), and on the other a proposal from industry for a method consistent with the way new processes are designed in the fine chemicals industry (Chapter 10). To me, a long-term stumbling block will be the integration of BPEO into the design process. The information required by HMIP in calculation of integrated environmental indices would normally only be available after substantial effort has been spent on design of a given option. The results of rigidly requiring such calculations would be high design costs to industry, with substantial effort spent on designing equipment which is not ultimately built. If, however, data from the early stages of design are used for comparison, the values are likely to be so approximate as to render comparison meaningless.

Other thorny issues include the question of how much of a product life cycle falls within the scope of the manufacturer's environmental responsibility, how much information should be disclosed to the public — and in what form — and how risks of releases (rather than actual releases) should be dealt with.

One might argue that, given all of the problems, implementation of an EMS standard (BS7750, for example) is not likely to be cost effective except as a public relations or marketing gimmick. However, pressure to implement standards is likely to grow. One emerging issue is the likelihood that regulators will move towards the use of a 'systems audit' approach — in other words, ensuring that companies have appropriate systems (both management and hardware) in place to control their environmental perfomance. In this context, the presence of a well organized, well documented and independently certified system is likely to bring benefits to both regulator and operator. A further key factor is the possibility of greater public credibility.

The solutions which are adopted to meet the challenges of EMS are still in development. Industry, standards organizations and the regulators are feeling their way. I hope this book will prove useful in the development process.

Finally, it is important to recognize the efforts of those who have devoted their time to the organization of the two-day IChemE North Western Branch conference held at UMIST, Manchester on 26 and 27 April 1995, on which this book is based. I would like to thank Ian McConvey (organizer of the conference programme), Tony Thompson (conference organizer), the session chairmen David Shillito and Tim White, and all of the authors for their invaluable contributions.

Paul Sharratt

THE AUTHORS

1. Managing public expectations and information needs
   Judith Petts *(Centre for Hazard and Risk Management, Loughborough University of Technology)*

2. Catchment management plans — linking standards, the water environment and people
   Craig Woolhouse *(National Rivers Authority)*

3. Lessons from the latest batch of HMIP prohibition notices
   Michael Sparshott *(Environmental Consultant)*

4. BS7750 and certification — a developmental history
   Christopher Sheldon *(British Standards Institution)*

5. The development of international standards
   David Hunt and Catherine Johnson *(WRc alert)*

6. Survey of industrial experience with environmental management
   Cynthia Riemann and Paul Sharratt *(Department of Chemical Engineering, UMIST)*

7. Making environmental management work
   Stuart Page *(Courtaulds Fibres)*

8. Developing a system to meet BS7750 at a manufacturing site
   Pauline Fawcett and John Nobbs *(BNFL)*

9. A practical approach to implementing BS7750 in the chemical industry
   Ken Jordan *(Akzo Nobel Chemicals)*

CONTENTS

1. MANAGING PUBLIC EXPECTATIONS AND INFORMATION NEEDS

Judith Petts

The UK Department of the Environment's 1993 survey of public attitudes to the environment revealed that public concern for the environment had not diminished during the recession and was almost at the peak level of the 1980s[1]. At the same time there have been reported rises in numbers of public complaints about industrial processes regulated by Her Majesty's Inspectorate of Pollution (HMIP) and about noise and odour pollution as reported by the Institute of Environmental Health Officers[2,3]. In the United States, national opinion polls have identified that public concerns about the environment do not appear to equate with governmental priorities for reducing risks[4].

PUBLIC CONCERN

The US studies point to a need for greater understanding of the reasons why policy-makers, industry and the public view environmental problems differently, as it is only with such understanding that acceptable management policies will be developed. Government agencies and industry are placing greater emphasis upon the need for better environmental communication — for example, the role of environmental reporting being stressed in the European Fifth Action Programme on the Environment[5], the EC Audit Regulation[6], 'Agenda 21'[7] and in various industry and business codes, charters and agreements (for example, the chemical industry 'Responsible Care' programme and the Business Charter for Sustainable Development of the International Chamber of Commerce). However, even with the best-intentioned efforts to provide the public with information about environmental impacts and risks, industry's credibility gap does not seem to be improving[8].

One of the primary reasons for this situation is that public concerns are not based simply upon a lack of information. Primary explanations for public concerns lie in:

* a loss of trust in governments and industry as credible sources of information and in their ability to take appropriate action to minimize environmental risks;

- a failure of current decision-making systems to involve the public effectively in decisions which affect the environment.

This chapter seeks to elucidate the underlying bases of public concerns and perceptions. Throughout there is reference to perceptions of *risk*, with risk broadly defined to mean a 'harmful outcome'. It is clear that environmental problems are viewed as risks. A broader understanding of public concerns should provide a means of identifying information and other requirements for achieving consensus in environmental management, and of promoting the potential effectiveness of industry's information provision activities.

PUBLIC PERCEPTIONS

WHO ARE 'THE PUBLIC'?
It is easy to characterize everyone other than industry or the regulatory/statutory authorities as 'the public': not least it provides for ease of expression. However, this suggests a uniformity of group, interest, knowledge and concern that is rarely, if ever, apparent. A key element of effective communication is the identification of the relevant 'stakeholders' — those people who are likely to have an interest (not necessarily directly expressing an interest) in the activities and effects of the organization. Employees, neighbours, the local community, environmental and other interest groups, customers and consumers, shareholders, banking and financial interests and the media may all be relevant stakeholders. There are differences in knowledge, concerns, agenda and communication needs between these groups, and indeed there are differences within each group.

What is usually evident — for example, in a local community close to an industrial site — is a spectrum of interests, which may appear to have a common theme when articulated (for example, concern about noise from a plant, or about air emissions) but which are actually underpinned by a complex array of perceptions, experiences and concerns. A tendency to believe that all stakeholders have the same agenda, and hence information needs, can lead to a failure to communicate effectively[9].

A view that stakeholders can be stereotyped in the way that they will respond to environmental issues is just one of the common fallacies about perceptions. Others include the view that 'the public' perceive risks while scientists deal in 'real' risks; that people will accept in the future risks which they have apparently accepted in the past; and that more rigorous approaches to identifying and controlling environmental problems (inherent, for example, in the adoption of environmental assessment and environmental management approaches) will provide some comfort to the public[10]. In relation to attempts to explain Not-In-My-BackYard (NIMBY) responses there has been evidence amongst industry, and sometimes politicians, of derogatory and negative views seeing public concern as merely being based on self-interest and/or irrational fears and that education of the public is the required remedying tactic[11].

THE PSYCHOLOGICAL DIMENSION OF PUBLIC PERCEPTIONS

The public perception literature is large and has a history going back at least 25 years. Reviews of research into perception (for example, References 12 and 13) identify four themes:
- the objective/subjective debate;
- the psychology of risk perception;
- social and cultural approaches;
- risk communication.

Expressed in this order these four themes indicate the gradual development and maturity of understanding of public perceptions. Understanding runs from the simplistic and now discredited view that risk can be considered in terms of either an assessed risk or a perceived risk, through to recognition that the influences upon public reactions to risks represent a complex interplay between judgements not only of the physical characteristics and consequences of an activity, but also social and institutional factors. Particularly important amongst the latter are issues such as trust in risk managers and decision-makers, concern over inequitable siting, lack of personal control, the pace of technological diffusion into social and cultural environments, and the extent to which the public can influence, and be involved in, decisions about environmental risks[14].

Psychometric methods have been employed in a large number of studies both in the US and Europe to explore the qualitative characteristics of risk. Psychometric techniques have been used to identify the similarities and differences among groups with regard to perceptions and attitudes. While some of the results in themselves must be evaluated with caution, the following variables have been found commonly to affect the perceived seriousness of risks:
- the expected number of fatalities or losses;
- the catastrophy potential;
- qualitative risk characteristics (for example, dread, control, blame);
- the beliefs associated with the cause of the risks.

The cultural view of risk is that people select what, and how much, to fear largely as the product of a particular cultural bias or to support a particular view of life. The perceiver is rarely an isolated individual, but exists within networks of social arrangements or institutions and information systems[15]. The cultural/social literature not only identifies the different values and concerns of men and women, the old and young, and the influences of class and education, but also the strong influence of the cultural and social setting in which people live and where information is gained, including the influence of the media[15,16].

TRUST AND CREDIBILITY

Trust and credibility have come to be seen as key determinants of perceptions and acceptability of risk within these networks — specifically the extent to which individuals trust risk managers, with trust being measured in relation to competence, ethics and intentions[17]. Trust has been seen to have relevance to a range of societal risks, but has gained a particularly high profile in North American studies of opposition to the siting of LULUs (Locally Unacceptable Land Uses) from the mid-1980s until the present day[18]. Trust relates to two different but overlapping factors:

- the individual or source providing the information;
- the institution or organization.

In the US context, Laird suggests that lack of trust in institutions may be symptomatic of a more general loss of faith in institutional arrangements and an unwillingness by the public to delegate responsibility for decisions to institutions and organizations which have regularly been seen to fail in their responsibilities[19]. This loss of faith can be witnessed in demands for public involvement in decision-making. Such demands are likely to present a major challenge over the next decade to decision-making procedures which are currently based upon relatively passive consultation procedures.

In the UK, the problem of a lack of trust has been seen to involve a number of components[10,20]:

- a lack of trust in industry to consider safety and the environment as seriously as making a profit;
- a lack of trust in fragmentary policy and regulatory systems to produce any co-ordinated and coherent strategies for managing risks;
- a lack of trust in different regulatory agencies and plant operators to monitor facilities effectively;

4

- highly visible expert disagreements about the reliability and validity of environmental risk assessments;
- the apparent lack of co-ordination between different authorities and the perceived artificial nature of divisions of responsibilities between different regulatory authorities;
- the fact that officials and managers in industry have, in the past at least, lacked adequate training in community and media relations and in the specific requirements of environmental risk communication.

Trust and confidence in industry and governmental sources of information derive from perceptions of competence and expertise, honesty and openness, and dedication and commitment. Perceptions of competence and expertise are largely influenced by an organization's environmental record, and by the spokesperson's merit factors such as track record, experience, presentation skills, education, professional recognition and independence. In the US, research has attempted to identify 'who' the public trusts to provide environmental information. While doctors, academics, non-profit organizations and environmental groups are seen as respected, informed and trustworthy, environmental consultants, industry and governmental representatives are seen as untrustworthy, biased and not caring[17]. Unfortunately, as yet the research results have not been tested in the UK cultural and institutional context. One of the particularly interesting findings of the US research, however, is the low level of trust attributed to environmental consultants who are often chosen by industry to undertake 'independent' assessments and audits on its behalf.

COMMUNICATION AND PUBLIC INFORMATION OBJECTIVES
Recognition of the need to communicate with the public is a relatively recent phenomenon. Public consultation requirements have underpinned certain regulatory activities, most particularly land-use planning, and the concept of the public 'right-to-know' has supported open access to information in some countries. However, the view that industry and businesses should communicate directly with the public and open a dialogue with them has only gained general credence in the last decade.

The new concern with informing the public can be seen to have several motivating sources, not entirely consistent with each other depending upon their focus (that is, industry or government). The motivating sources include:
- a desire to improve image (ultimately to commercial advantage);

5

- a desire to overcome opposition to decisions;
- a desire to pre-empt or preclude the potential for opposition to develop;
- a desire to develop effective alternatives to direct regulatory control;
- the *right* of the public to know about health, safety and environmental risks which may affect them;
- a desire to share power (as a means of reducing conflict).

Hence, several reasons for communicating can be identified[10]:

- reassurance — for example, of the ability of an industrial operator to detect hazardous conditions; of the ability to prevent accidents; of the degree of understanding and competence of the operator; of compliance with legislation; of effective performance;
- persuasion — for example, of the appropriateness of the site chosen for a new development; of the appropriateness of the proposed risk benefit trade-off implicit in the proposed decision;
- arousal — for example, provision of emergency action information to the public; safety information given to the work force.

The provision of information, and specifically industrial 'reporting', can be mandatory or voluntary and undertaken at the level of the individual firm or industry sector or at the site or corporate level. The question is what is the appropriate balance to achieve between mandatory and voluntary reporting.

The principle of the 'community right-to-know' underpins many institutional communication objectives. In the US, the SARA Title III requirements and in Europe, Directive 82/501/EEC, provide examples. In the 'right-to-know' scenario communication has to be formalized because the normal informal communication networks which exist in any social context are inadequate or not assured — that is, industry may not naturally communicate with its neighbours, and it is necessary to ensure that all parties who might be affected are given an equal opportunity to participate and to take action based on a common set of information.

In the context of the company environmental report, the provision of information is seen as an issue of responsibility, accountability and sustainability. In relation to the latter, US experience suggests that companies that are forced — or encouraged — to publish sensitive performance data will soon look to emission reduction programmes, a trend accelerated by the activities of public pressure groups and ethical investment groups who have increasingly used the published data to benchmark the environmental performance of companies[21]. The 1992 UN Conference on Environment and Development (UNCED) commitment to sustainability (Agenda 21) requires, and depends upon for its ultimate success, increased reporting by industry. Information is to form the basis of response and action whether by governments, industry or the public.

The community right-to-know ethic can be seen as a means of empowering citizens to participate in the policy process because it removes, at least in part, one barrier to participation — that is, lack of relevant information. It forms the base of the 'ladder' of participation[22], with formal consultation and proactive involvement in influencing decisions forming the next 'rungs' up the ladder.

Reflecting this theme of the empowering role of information, the US National Research Council has defined risk communication as 'an interactive process of exchange of information and opinion among individuals, groups and institutions'[23] — that is, communication becomes more than a passive one-way provision of information from industry or government to interested stakeholders. There is recognition in this definition of the need for a two-way flow of information. The combining of opinions with information places considerable demands upon the communication process and perhaps this is easier to imagine where the objectives of communication are highly proactive, such as in relation to the objective of persuasion relative to the siting of a new facility, rather than where it is perhaps the more passive communication scenario of producing an annual environmental report. The latter has to be seen in the much broader context of communication, being part of the whole communication strategy of an organization, and an activity which in itself will impact upon the success of other communication activities at other times.

Communication as a two-way process has significant procedural implications requiring different methods (discussion groups, advisory fora, liaison groups, etc) and timing than the simple provision of information. Consensus-building and conflict-resolution techniques have been advocated and tested, particularly for siting decisions at the local level in North America[24,25]. Citizen panels have been advocated for use in topics ranging from city planning to long-range energy planning[26]. In the UK some local authorities are starting to use consensus-building techniques in plan-making functions[27]. Similar pressures have begun to be seen in relation to industrial activity, with the local liaison group perhaps providing the most common example, and the appointment of a local community representative to the board of a company (as in the case of South East London Combined Heat and Power) perhaps the least common.

MEETING PUBLIC INFORMATION AND COMMUNICATION NEEDS EFFECTIVELY

From all this it should be apparent that, within the context of environmental management systems, effective environmental and risk communication requires more than the provision of performance information in glossy annual reports and the publication of audit results. It lies in genuine efforts to provide for the public to understand the day-to-day activities of industrial operations and also in real, measurable and visible improvements in environmental performance. Furthermore, industry alone is not able to meet all of the public's information needs, and the extension of public involvement in decision-making processes is also required.

IMPROVING INDUSTRIAL PERFORMANCE

An annual environmental report which shows a decrease in sulphur dioxide emissions or in solid waste disposal from an installation will quickly be undermined by continuing problems with odour or noise. The power of the media with regard to highlighting inappropriate and unsuccessful control cannot be underestimated. Control of facilities by good day-to-day management, effective facility monitoring, and potential for local communities in particular to have open access to plant and to monitoring output, to liaison group meetings and so on, is fundamental to improving perceptions, although is not going to be achieved in a short time period.

One of the biggest problems in any industrial sector is the influence of the 'worst performer' (in environmental terms) on the public perception of the best performer. Here the role of sector-wide codes and guides has an important role to play in the context of self-regulation, although strong and effective *direct* regulation will remain fundamental to control and also to the promotion of public trust.

In the context of sustainability, companies can expect to come under pressure to not only perform better in terms of reducing emissions. They must also acknowledge their share of the responsibility for the depletion of non-renewable resources and contribute to the development of a more sustainable lifestyle[8]. Within the environmental management system, solid performance and tangible evidence of reductions in environmental impact are essential to build confidence in a company's commitment and stewardship.

IMPROVING RISK INFORMATION AND UNDERSTANDING

At the heart of public opposition is mistrust of the extent to which industry and regulators have adequately and rigorously assessed existing and potential environmental impacts (including risk to health) and understand the nature of the impacts. Particular problems for environmental reporting appear to arise in:

- understanding and measuring a full inventory of emissions release across an entire business cycle — for example, for power generation the full cycle of exploration, extraction, processing, distribution, combustion and power supply;
- understanding and measuring the impacts of indirect upstream and downstream activities;
- relating emissions to actual damage.

There is a general need to improve the quality and availability of data and information on the environmental impacts of industrial activities. This need extends beyond the operations of specific plant to the general research, regulatory and expert community. For example, 'expert' understanding that dioxins from industrial activities are not a health problem has to be *proven* to the public by sound monitoring and auditing of facilities and rapid response to dealing with concerns that arise in relation to specific sites. This information also has to be in the public domain. There is greater scope for industry to support third party research and audits.

There needs to be attention to the use and development of structured and objective risk assessment and risk management approaches to environmental problems, although a primary problem remains how to incorporate cost-benefit considerations effectively into these assessments. There is a need to ensure that such structured assessment approaches underpin all stages of activity, from plant design, through siting, operation, auditing, decommissioning and dealing with closed and old sites. Risk assessors have to be open about data inputs to assessments, the assessment and modelling methods used and the uncertainty in risk estimates. They must involve the public more effectively in decisions about the acceptability of risks, the latter ultimately being a question of values.

In a study of environmental reporting by 100 companies[21], five stages in corporate reporting are discussed. They start at a basic level of 'green' publicity brochures and videos, through the stage of annual reporting linked to an environmental management system, to the top level of 'sustainable development' reporting which will be based on the extensive use of quantitative methods (such as life cycle assessment and mass balances), and on strong links with

9

industry-wide and national sustainable development reporting against pre-agreed targets. This stage 5 is predicted to become the 'next goal' of the business community as yet virtually untouched.

IMPROVING COMMUNICATION

Trust and credibility cannot be built quickly. They are the result of ongoing partnerships, actions, performance and skill in communications. Perceptions of commitment and diligence in the pursuit of environmental goals are gauged by all of the actions, and verbal and non-verbal communications of managers and representatives of an organization. What might seem like very small issues to the company — such as making a representative available to answer questions at a local meeting, and that person being prepared to stay late or to leave a phone number for further contact — can have a considerable impact on the view that people hold of an organization. Caring, empathy, competence, honesty, openness and expertise are the key requirements of communicators in terms of public credibility. Research indicates that the first three of these requirements are in fact a greater influence on credibility than expertise, and that even the gender of the communicator has a pronounced effect on these dimensions of credibility[17].

In the US, the Environmental Protection Agency has published seven rules or guidelines for effective communication by officers[28] and has initiated a programme of enhanced community relations and extensive field testing of communications materials. Communication is a management skill, which requires training and development beyond simple attendance on a media training course.

One of the few empirical studies of risk communication which has been undertaken is research for the Commission of the European Communities in relation to implementation of Article 8 of Directive 82/501/EEC, which requires information to be given to the local community in the vicinity of major hazard installations[29]. The study concludes that even where the provision of information is a regulatory requirement, effectiveness requires more than just information. It needs a commitment to developing communication between all parties over a prolonged period. This requires a strategic framework of developing social relations as the context within which information strategies are designed and negotiated. Understanding of the social context and the factors which influence the development of perceptions and concerns and the consequent information requirements is important.

For any individual company understanding 'who' its key stakeholders are (including the company's own employees as both interested parties and important communicators) is essential within a management system. Sometimes, companies are known to feel that local community interest in their activities is low, and that there is even a certain degree of apathy amongst neighbours with whom it tries to communicate[8]. One study in the US concluded that neighbours appear to be less interested in information and emissions data than might be expected, but the complacency was partly explained by the sheer volume of information being presented and its relative novelty. Such situations stress the importance of providing a number of means of communication — for example, for environmental reports to provide opportunities for direct contact with the company. Companies need to provide information to the different stakeholders which can be absorbed by those to whom it is directed and which takes into account the different values, needs and interests of different groups. Stakeholder analysis to understand the latter should be an important component of a management system. This in turn leads to identification of different communication tools of relevance to the different groups — that is, reports, press briefings, open house events, door-to-door mailings, posted signs, manned information points, material and product information sheets and so on.

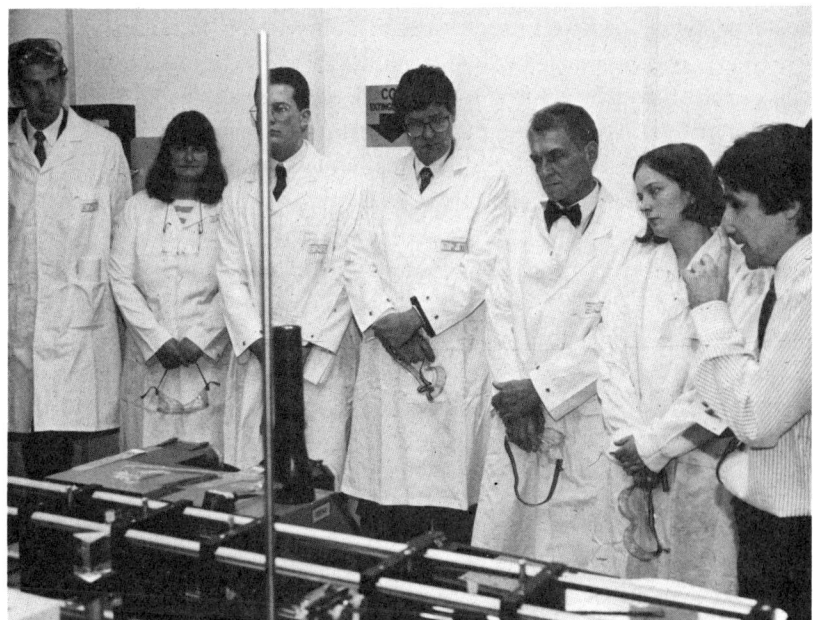

An effective communication tool — the company open day. (Courtesy of BNFL.)

CRITERIA OF EFFECTIVE COMMUNICATION

Within the context of environmental management systems, it seems important to consider criteria for assessing the effectiveness of communication and information provision. At least four criteria might be considered:

- the quality of the relationship between an industrial installation and the local community, not least the level of trust in the company;
- the effects of public information upon management within an organization;
- the general information climate that exists relating to the management and regulation of industry;
- the general extent of 'empowerment' of the public — that is, helping people to achieve their own purpose by increasing their confidence and capacity.

CONCLUSIONS

Public concerns about industrial environmental performance and management are extremely complex and not simply a reflection of information availability but also of trust. Understanding and accepting these concerns as valid and rational is essential. Building trust and credibility once it has been lost is going to be a lengthy and perhaps in some aspects an impossible task. Environmental reporting is one very small but important tool to help to begin to redress credibility. If the reporting is found to be meaningless or superficial, however, its potential will be lost. The whole environmental management system (of which the report may form a part) requires attention to the setting of targets against which performance can be effectively measured and publicly proven, analysis and understanding of stakeholder information needs, and training of effective communicators. But industry cannot address public concerns by itself; the role of environmental experts, regulators and governments is also important. In particular, information will have to be backed by attention to improving the means by which the public can be involved, rather than merely consulted, in environmental management decision-making.

REFERENCES IN CHAPTER 1

1. Department of the Environment, 1994, *Digest of Environmental Protection and Water Statistics*, No. 16 (HMSO, London, UK).
2. Her Majesty's Inspectorate of Pollution, 1994, *7th Annual Report* (HMSO, London, UK).

3. Anon, 1993, EHOs' report shows upturn in noise and odour pollution, *ENDS*, 224: 8–9.
4. United States Environmental Protection Agency Science Advisory Board, 1990, *Reducing Risk: Setting Priorities and Strategies for Environmental Protection* (USEPA Science Advisory Board, Washington DC, USA).
5. Commission of the European Communities, 1992, *Towards Sustainability — A European Community Programme of Policy and Action in Relation to the Environment and Sustainable Development,* COM(92)23 Final — Vol. II (CEC, Brussels, Belgium).
6. Commission of the European Communities, 1993, Council Regulation allowing voluntary participation by companies in the industrial sector in a Community eco-management and audit scheme, EEC No. 1836/93, *Official Journal of the European Communities,* No. L 168/1, 10.7.93.
7. United Nations Conference on Environment and Development, 1992, Rio Declaration on Environment and Development.
8. United Nations Environment Programme Industry and Environment Office, 1991, *Companies' Organization and Public Communication on Environmental Issues,* Technical Report Series No 6 (UNEP, Paris, France).
9. Petts, J., 1994, Effective waste management: understanding and dealing with public concerns, *Waste Management and Research,* 12: 207–222.
10. Petts, J., 1994, Risk communication and environmental risk assessment, *Nuclear Energy,* 33 (2): 95–102.
11. Wolsink, M., 1994, Entanglement of interests and motives: assumptions behind the NIMBY-theory on facility siting, *Urban Studies,* 31 (6): 851–866.
12. The Royal Society, 1992, *Risk: Analysis, Perception and Management* (The Royal Society, London, UK).
13. Krimsky, S. and Golding, D. (eds), 1992, *Social Theories of Risk* (Praeger, Connecticut, USA).
14. Renn, O., 1992, Risk communication: towards a rational discourse with the public, *Journal of Hazardous Materials,* 29: 465–519.
15. Douglas, M. and Wildavsky, A., 1982, *Risk and Culture* (University of California Press, Berkeley, USA).
16. Thompson, M., Ellis, R. and Wildavsky, A., 1990, *Cultural Theory* (Westview Press, Boulder, Colorado, USA).
17. Covello, V.T., 1992, Risk communication: a new and emerging area of communication research, in *Risk Assessment, Proceedings of a Conference on Risk Assessment, 6–9 October, 1992* (Health and Safety Executive, Bootle, Merseyside, UK).
18. Portney, K.E., 1991, *Siting Hazardous Waste Facilities* (Auburn House, New York, USA).
19. Laird, F.N., 1989, The decline of deference: the political context of risk communication, *Risk Analysis,* 9: 543–550.

20. Petts, J., 1992, Incineration risk perceptions and public concern: experience in the UK improving risk communication, *Waste Management and Research,* 10: 169–183.

21. United Nations Environment Programme Industry and Environment Office, 1994, *Company Environmental Reporting,* Technical Report No 24 (UNEP, Paris, France).

22. Arnstein, S., 1969, A ladder of citizen participation, *Journal of the American Institution of Planners,* 35: 216–224.

23. National Research Council, 1989, *Improving Risk Communication* (National Academy Press, Washington DC, USA).

24. Bacow, L.S. and Wheeler, M., 1984, *Environmental Dispute Resolution* (Plenum Press, New York, USA).

25. Susskind, L. and Cruikshank, J., 1987, *Breaking the Impasse: Consensual Approaches to Resolving Disputes* (Basic Books, New York, USA).

26. Renn, O., Stegelmann, G., Albrecht, U. Kotte and Peters, H.P., 1984, An empirical investigation of citizens' preferences among four energy scenarios, *Technological Forecasting and Social Change,* 26: 11–46.

27. Petts, J., 1994, *Hampshire County Council Integrated Waste Strategy Community Consultation and Involvement: A Case Study,* Energy Technology Support Unit Report No. B/EW/00389/23/REP (ETSU, Harwell, UK).

28. Covello, V.T. and Allen, F., 1987, *Seven Cardinal Rules of Risk Communication* (US Environmental Protection Agency, Office of Policy Analysis, Washington DC, USA).

29. Wynne, B., 1992, *Empirical Evaluation of Public Information on Major Industrial Accident Hazards,* EUR 14443 EN (Joint Research Centre Commission of the European Communities, CEC, Luxembourg).

2. CATCHMENT MANAGEMENT PLANS — LINKING STANDARDS, THE WATER ENVIRONMENT AND PEOPLE

Craig Woolhouse

The National Rivers Authority (NRA) was established in 1989 as the 'guardian of the water environment' in England and Wales. Under the Water Resources Act 1991 it has statutory duties and powers in relation to water resources, pollution control, flood defences, fisheries, conservation, recreation and navigation. The water quality and fisheries responsibilities extend into coastal waters. The NRA is the Competent Authority for some twenty European Community Environmental Directives.

The NRA is a non-departmental public body with policy links to the Department of the Environment (DoE), the Ministry of Agriculture, Fisheries and Food (MAFF) and the Welsh Office. The Authority has head offices in Bristol and London and operates through eight regions and 26 areas. The head offices are primarily concerned with policy development and performance monitoring, and the regions and areas with policy implementation and day-to-day operations. The NRA has an annual budget of £450M and 7500 staff.

INTEGRATED MANAGEMENT OF THE WATER ENVIRONMENT

Over the last three years the NRA has created 26 operational areas. The purpose of these areas is to deliver integrated services in order to balance the needs of water users with those of the environment. In addition to delivering multifunctional operational activity, the areas are also responsible for integrated planning in order to achieve a sustainable water environment. Catchment management plans (CMPs) are the means of delivering this activity.

The process of planning for environmental sustainability and improvement through an integrated approach to river catchment management is not an inward looking process. The aim is to use proactive planning to prevent future environmental damage through active involvement with the public and private sectors.

This chapter explores the process adopted by the NRA to deliver integrated management of the water environment, identifies the links between CMPs and other environmental management systems and describes some recent

NRA initiatives which are reflected in CMPs. It concludes by looking forward to the creation of the UK Environment Agency.

CATCHMENT MANAGEMENT PLANS

THE PROCESS

Catchment management planning[1] is the process by which the problems and opportunities resulting from catchment uses are assessed and action is proposed to optimize the overall future well-being of the water environment. A catchment use is defined as a direct use of the water environment (for example, water abstraction) or an activity which impacts upon it (for example, mineral extraction). Catchments are defined as discrete geographical units with boundaries derived primarily from surface water considerations and comprise one or more hydrometric sub-catchments.

The approach to producing CMPs is based on many of the accepted building blocks for the integrated planning of natural resources (for example, community participation, consideration of physical characteristics, shared policy development and action, and strategic analysis of issues).

The plans consider the various water users' interests and develop a long-term vision and medium-term strategies and actions through consultations with local communities and organizations. Each plan is produced within a twelve to eighteen month period. The initial step is to prepare a consultation report. After a period of open consultation an action plan is then produced.

The purpose of the consultation report is to describe the resources, uses and activities relevant to the water environment, set environmental objectives (in relation to water quality and the physical environment), present issues and suggest potential solutions, and promote a catchment vision. Public consultation then takes place over a period of eight to thirteen weeks through public meetings, press releases and direct mailing.

The action plan subsequently sets down the agreed vision, strategy and activity plans for the catchment. It reflects the results of the consultation process. The vision looks to the long term (ten years plus) whereas the strategy and activity plans have a three to five year time horizon. Annual monitoring of commitments is undertaken and reported by the NRA. A range of organizations and groups are likely to be responsible for undertaking actions. Informal liaison with key partners (for example, local authorities, water companies and conservation groups) is an essential element of the planning process.

IMPLEMENTATION OF CMP

Preparation of plans is a shared activity between area and regional staff but the responsibility for implementing the plans lies with the 26 NRA multifunctional operational areas. Each area, operating within national policy guidance, has the responsibility across all the NRA areas of activity to identify the key issues facing the local water environment and to tackle them directly or in partnership with other organizations. By January 1995 the NRA had produced 63 consultation reports and 20 action plans (see Figure 2.1 on page 18). A total of 163 CMPs will be produced by the end of 1998.

The NRA has learnt a number of key lessons through the process of producing plans[2]:

- the planning process must be as short as possible;
- planning on the basis of what is already known is an acceptable risk provided subsequent action is subject to detailed evaluation;
- integrated planning is feasible but all-round management support is needed to translate it into practice;
- the process, rather than the size of the catchment, determines overall resource needs. Minor catchments should be grouped to minimize costs;
- plans should not raise expectations beyond what is achievable. They should, however, include a long-term vision for the catchment;
- local communities respond very positively to being involved in the planning process;
- inter-functional co-operation and understanding within the NRA is improved;
- catchments have their own particular needs but a nationally consistent approach has to be fostered;
- further research is required to develop objectives and standards.

ENVIRONMENTAL MANAGEMENT SYSTEMS

The public and private sectors are pursuing a wide range of initiatives (statutory and voluntary) whose overall aim is the better management and protection of natural resources. Economic competitiveness and social considerations are inextricably linked to these initiatives. Activities range from those intended to improve the particular operations of a company (for example, BS7750) to strategies aimed to make lifestyles more compatible with environmental values (for example, UK Strategy on Sustainable Development[3]).

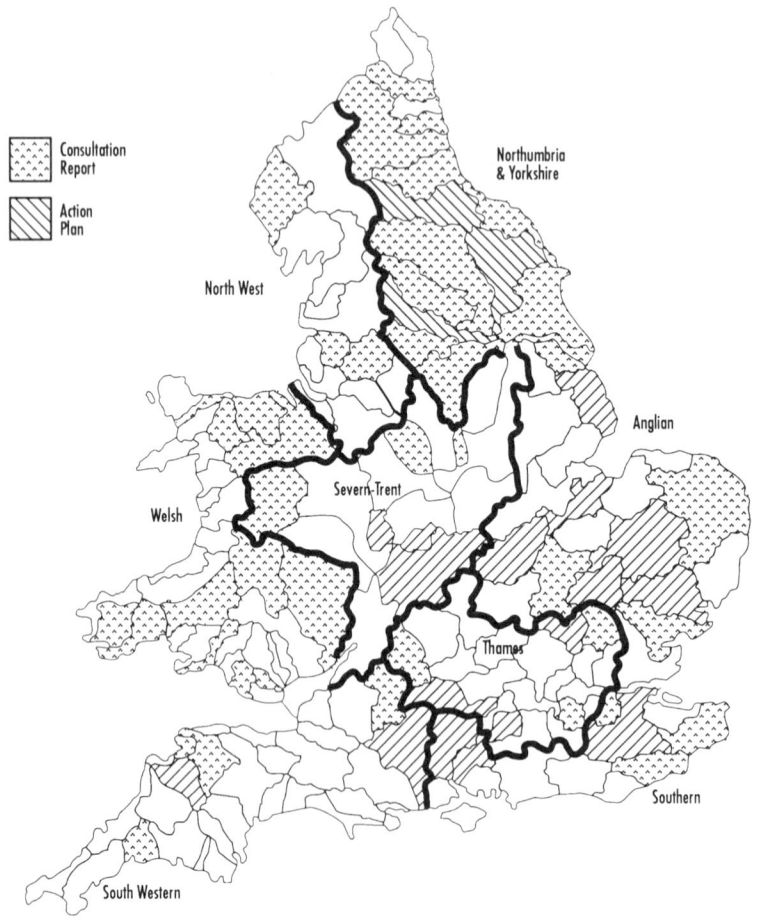

Figure 2.1   Availability of catchment management plans.

The debate on the wise use of natural resources received added impetus at the United Nations Conference on Environment and Development, held in Rio de Janiero in June 1992. One of the key outputs from that conference was Agenda 21, described as 'an action plan for the 1990s and 21st century, elaborating strategies and integrated programme measures to halt and reverse the

Figure 2.2   Environmental management systems — context and links.

effects of environmental degradation and to promote environmentally sound and sustainable development in all countries'[4].

A number of themes are discernible in Agenda 21:

• the need to integrate environmental management systems (EMS) — both planning and operational — across traditional sectoral boundaries;

• the need to involve communities in environmental decision-making;

• the need to develop appropriate EMS for different sectors and purposes within a strategic framework.

The principles of sustainable development should embrace all EMS and provide a common basis for ensuring that policies, legislative controls, techniques and strategies are complementary and inter-linked (see Figure 2.2). CMPs have a role to play in influencing other strategies and more site-specific initiatives. In addition they provide an opportunity for local communities to influence the environmental management framework for their local water environment.

CMPS AND OTHER ENVIRONMENTAL MANAGEMENT SYSTEMS

Four examples of EMS are now described and their relationship with CMPs highlighted.

Integrated pollution control (IPC)
Embodied in the Environmental Protection Act 1990, IPC provides a mechanism for looking at the impact which an industrial process has on the whole environment. The objectives of the approach are to:
- use the best available techniques not entailing excessive cost (BATNEEC) for preventing pollution;
- ensure BATNEEC is used to minimize overall pollution by following the best practicable environmental option (BPEO);
- comply with all relevant environmental standards.

For all IPC discharges to water, the overall river quality planning targets set by the NRA are used to evaluate the potential of the river to receive additional effluent. The planning targets reflect European Community Directives (for example, 78/659/EEC: The Quality of Fresh Waters Needing Protection or Improvement to Support Fish Life) and river use (for example, River Ecosystem) related standards. The role of CMPs is to consult informally on the current and future uses of rivers so that appropriate river quality planning targets can be set. In particular, the potential for long-term (that is, post 2005) planning targets to be tightened is being considered, so that the benefits and costs of potential action can be assessed. The targets will reflect conservation, amenity, effluent discharge and recreation uses of a river stretch.

Land-use development plans
Land-use development plans are given statutory force by the Town and Country Planning Act 1990 and provide for the optimum use of land for urban development subject to the needs of local people and the environment. Planning Policy Guidance Note 23 published in June 1994 recognizes that local land-use policy and development control decisions can play a key role in minimizing pollution risk in conjunction with the statutory controls of the environmental regulators. Development plans are invariably examined at public inquiry.

Through CMPs the NRA is preparing 'land-use statements' which integrate land- and water-use issues for the particular area being considered. The range of issues[5] to be covered includes:
- the constraints on development that may result from a lack of water resources or sewage treatment capacity;
- the need to protect flood plains from development;
- the requirements to protect groundwater quality from redevelopment of contaminated land or waste disposal.

Environmental technology best practice programme
Launched by the UK Government in June 1994, the purpose of this programme is to promote better environmental performance by industry, primarily through waste minimization and cost-effective cleaner production processes. The Aire and Calder waste minimization project has shown how environmentally and economically successful such initiatives can be.

In opening a debate on the future health of the local water environment, the CMP may act as a catalyst for creating partnerships which can draw more effectively on central government or European initiatives to improve the water environment. The partnership model is becoming increasingly necessary as competitive bidding for specific funds, often to short timescales, is becoming more prevalent (for example, Single Regeneration Budget, EC LIFE Fund).

BS7750
This British Standard is an environmental management system for companies and organizations. It defines the parameters for establishing a regime for all aspects of a company's activity in order to ensure the most advantageous outcome for the environment.

By reviewing all the information relevant to the future management of a river catchment and establishing a long-term vision in conjunction with the local community, the CMP is an ideal reference point for local companies preparing BS7750-based management systems. The CMP will not only cover the legal standards but also review those areas in which environmental protection and enhancement should be achieved, for those wishing to deliver total commitment to the environment.

RECENT NRA INITIATIVES
The following three examples have been chosen to illustrate how recent NRA initiatives are guiding the production of CMPs as well as being implemented through operational work.

STATUTORY WATER QUALITY OBJECTIVES (SWQO)
SWQO were introduced by Section 83 of the Water Resources Act 1991. Their purpose is to establish clear statutory water quality targets in the 44,000 km of Controlled Waters throughout England and Wales. They will replace the existing informal river quality objectives. The SWQO scheme will be use-related.

21

Five river uses are envisaged:
- river ecosystem;
- special ecosystem;
- abstraction for potable supply;
- agricultural abstraction;
- water sports.

The classification and standards for the river ecosystem use have been introduced. Details of the other four standards are still at the development or approval stage.

The DoE has not yet provided the NRA with a timetable for the introduction of SWQO, although an announcement was due in February 1995. In the interim the NRA will liaise informally on WQO for Controlled Waters through catchment management plans. These plans will review existing performance and look at the options for tightening standards in the long term (that is, post 2005). This date represents the start date for implementing the next phase of water industry investment.

GROUNDWATER PROTECTION — MANAGEMENT OF NITRATE

There are three EC Directives which have a bearing on the control of nitrate in the water environment. EC Directive 75/440/EEC (Surface Water Abstraction) sets a limit of nitrate in surface waters used for drinking water of 50 mg/l in 95% of samples. The NRA is the Competent Authority for this directive; of the 500 abstraction points across England and Wales only eight regularly fail the standard.

EC Directive 91/676/EEC (Nitrates from Agriculture) requires that 'nitrate vulnerable zones' (NVZs) are identified in catchments where the nitrate standards are exceeded or could be exceeded. Within designated areas codes of practice must be established to manage the application of nitrogen fertilisers and manures. The directive is currently being implemented and the NRA has played a role in defining the NVZs which are now subject to consultation by the DoE and MAFF. The total area proposed is 650,000 hectares. Final details are due to be published shortly.

The EC Directive 91/271/EEC (Urban Waste Water Directive) contains provisions for designating areas which are eutrophic because of discharges from sewage treatment works. None of the first tranche of areas announced in May 1994 was designated on the basis of nitrate enrichment. Investigations on inland and esturial waters are continuing to identify potential areas for the 1997 review of sites.

The CMP will identify the extent to which these directives apply or could apply within catchments and review the concerns of conservation groups and water companies about the impact that nitrate may have on those uses.

CITIZEN'S CHARTER

The NRA was awarded the Citizen's Charter in 1994. Its approach reflects six key principles, including 'information and openness'.

CMPs play a key role in NRA's communications with local communities, interest groups and other statutory agencies. They contain relevant data and information which is presented in a way which explains the wide range of pressures the water environment is under and the way in which these are, and will be, managed. Information contained in plans may include:

- lists of abstractors;
- prosecutions for pollution;
- areas of flood risk;
- areas of conservation interest.

THE ENVIRONMENT AGENCY

The merger of the Waste Regulation Authorities (WRA), Her Majesty's Inspectorate of Pollution (HMIP) and the NRA to create the Environment Agency is planned to occur in April 1996. This change will influence the future direction of pollution management across air, land and water and the delivery of integrated water management.

The Secretary of State for the Environment has indicated (DoE news release, 13 October 1994) that the Agency should:

- ' ... be integrated geographically so that interconnected systems like river catchments are considered as a whole ... ';
- ' ... develop a close relationship with the public, business and other organizations, including Local Authorities ... responsive to local concerns ... '.

CMPs are involved in delivering both these principles.

NATURAL RESOURCE PLANNING BY HMIP, WRA AND NRA

The spatial impact of some pollutants (for example, sulphur dioxide emissions) over specific areas is considered when IPC authorizations are prepared by HMIP. In respect of discharges to water, the pollution loads are related to the water quality planning targets set by the NRA and are therefore directly related to natural resource units. For land and air emissions polluting loads are sometimes managed in relation to administrative, rather than natural resource, units.

The preparation of plans for Air Quality Management Areas (for example, urban areas) is seen as a necessary step to ensure that air quality is managed in an integrated, rather than a piecemeal, way. The aim of these plans is to use the relevant powers and influence of central government, HMIP and local authorities to achieve specific environmental standards. The lead organization for such plans could be any of the key groups, although recent proposals by the DoE and Department of Transport identify local authorities as the responsible body.

WRA currently prepare Waste Disposal Plans in accordance with the Environmental Protection Act 1990. These detail what waste needs to be dealt with and the implications for the environment of different disposal means. The plans cannot resolve the spatial issues of disposal. The siting of incinerators or landfill sites is a land-use planning matter covered by Waste Local Plans prepared by the relevant planning authorities. The Waste Disposal Plans are not subject to examination at public inquiry.

By April 1996 the NRA will have completed two-thirds of its first generation CMP programme. The further development of these plans, successor plans to the Waste Disposal Plans and plans for Air Quality Management Areas will raise important questions for the Agency in terms of integrated pollution/water management and the way in which these issues are discussed with interested parties.

CONCLUSIONS

CMPs deliver integrated management of the water environment at the local level and provide a means for listening to and responding to local concerns. In facilitating the preparation of a CMP, the NRA seeks to enable a wide range of interested parties to act in partnership in achieving environmental improvements. These plans have been welcomed by local authorities, central government, interest groups and other statutory agencies. The NRA has found the plans to be of great benefit in ensuring that the synergies between operational and regulatory activity are exploited for the benefit of the water environment in individual catchments. The involvement of local communities and their representatives in the CMP process has been fundamental in its success.

Access to environmental data is only of general benefit if the data are presented in a way which offers a clear picture. CMPs provide a range of data in a way which tells a story to local interests. Environmental reporting is of most benefit if it is undertaken at a local as well as a national level.

Achievement of sustainable development principles and targets depends in part on the effective integration of policy, regulatory controls and operational activity across government agencies. CMPs can be considered as one of a number of management systems and techniques.

REFERENCES IN CHAPTER 2

1. National Rivers Authority, 1993, *Catchment Management Planning Guidelines* (NRA, Bristol, UK).
2. Woolhouse, C.H., 1994, Catchment management plans: current successes and future opportunities, *Integrated River Basin Development* (John Wiley and Sons, Chichester, UK), 463–474.
3. Department of the Environment, 1994, *UK Strategy on Sustainable Development* (HMSO, London, UK).
4. United Nations, 1992, *Agenda 21: Programme of Action for Sustainable Development* (United Nations, New York, USA).
5. National Rivers Authority, 1994, *Guidance Notes for Local Planning Authorities on the Methods of Protecting the Water Environment through Development Plans* (NRA, Bristol, UK).

3. LESSONS FROM THE LATEST BATCH OF HMIP PROHIBITION NOTICES

Michael Sparshott

The introduction of integrated pollution control (IPC) under Part 1 of the Environmental Protection Act 1990, and the subsequent issue by Her Majesty's Inspectorate of Pollution (HMIP) of process authorizations for prescribed processes have led to a considerable increase in the regulatory burden on process operators. They are subjected to detailed requirements for operating practices, data reporting and limitations on emissions to the various compartments of the environment, all of which are backed up by force of law. Infringements of these requirements can lead, at the least, to the issue of an Enforcement Notice by the Inspectorate, requiring correction of the transgression(s) by a set date. In some cases, infringements can lead to a prosecution with a substantial fine and costs, and occasionally to a Prohibition Notice requiring immediate cessation of those operations which contravene the authorization.

A survey of press releases generated by HMIP as a result of various actions during the period May to December 1994 shows a wide variation in the technical causes of the failures which led to the actions. The majority, however, can be assigned to a small number of basic failures of procedures at one or more stages of the process of applying for, establishing and operating according to the requirements of the process authorization. This chapter highlights those basic failures, and gives some brief guidance on their avoidance.

SURVEY OVERVIEW

The survey looked at 39 press releases reporting infringements of the Environmental Protection Act 1990 (EPA90), and seven press releases reporting infringements of the Health and Safety at Work Act 1974 (HSW74), during the period May to December 1994. The seven HSW74 press releases were included because they relate to processes which will be brought under IPC in the very near future, and are expected to have a similar impact on the process operators under the new regulatory regime. The total of 46 press releases refer to HMIP actions shown in Table 3.1 on page 28.

TABLE 3.1
Breakdown of the HMIP press releases considered in the survey

Prosecutions under EPA90	5
Prosecutions under HSW74	3
Subtotal	8
Enforcement Notices under EPA90	33
Improvement Notices under HSW74	3
Subtotal	36
Prohibition Notices under EPA90	1
Prohibition Notices under HSW74	1
Subtotal	2
Total	46

TABLE 3.2
The fines and costs imposed as a result of seven prosecutions

	Fine, £	Costs, £	Total, £
Under EPA90	10,000	5135	15,135
	5000	5000	10,000
	7500	6040	13,540
	5000	12,062	17,062
	22,500	10,000	32,500*
Under HSW74:	4000	1540	5540
	12,000	5400	17,400
Total (both Acts)	66,000	45,177	111,177

* The large fine of £22,500 resulted from a single prosecution covering 75 separate offences of three different types, arising from 25 separate incidents.

Of the seven prosecutions for which data were available, the fines and costs imposed by the magistrates' courts are given in Table 3.2.

It should be borne in mind that, substantial though some of the fines and costs are, they represent only the external costs of failures leading to a prosecution. To these should be added the following:

- the internal company costs of the action — for example, defence costs;
- the costs of resources required to correct the problem;
- the costs of possible losses of production;
- the unquantifiable costs of resultant adverse publicity.

TYPES OF FAILURE, AND THEIR FREQUENCY OF OCCURRENCE

The failures mentioned in the press releases as leading to the actions described have been extracted and ranked in order of frequency of occurrence (that is, the number of HMIP actions to which a given failure type has contributed). This includes both EPA90 and HSW74, and takes Improvement Notices (HSW74) as being equivalent to Enforcement Notices (EPA90). Since many of the HMIP actions were based on incidents which included more than one type of failure, a total of 83 failures are distributed among 17 failure types.

The failure types, ranked according to frequency of occurrence, are listed briefly here, and are discussed in detail in the next section. Where a failure type contributed to a prosecution or prohibition, this is indicated. Otherwise, only Enforcement or Improvement Notices were involved.

(a) Lack of (or inadequate) operational procedures to prevent or minimize emissions — thirteen failures, including three prosecutions and one prohibition.

(b) Failure to complete improvement programmes prescribed by the process authorization by the date set in the authorization — nine failures.

(c) Emission to air (leakage or venting) due to operational failure, or inadequate gas treating or scrubbing (for example, failure to use BATNEEC) — eight failures, including three prosecutions and one prohibition.

(d) Emission to air or water due to failure of maintenance or repair procedures — seven failures, including two prosecutions.

(e)   Inadequate control of discharge to water due to operational or equipment failure, or not using BATNEEC — six failures, including three prosecutions.

(f)   Contravention of operating conditions contained in a process authorization (plant or control system not in operation) — six failures.

(g)   Measurement of emissions to air inadequate in comparison with conditions contained in a process authorization — five failures.

(h)   Inadequate procedures to ensure compliance reporting required by the conditions of a process authorization — five failures, including one prosecution.

(i)   Failure to notify HMIP when a condition of a process authorization is exceeded (for example, emission of a component above a concentration limit) within 24 hours following the emission — five failures, including one prosecution.

(j)   Lack of (or inadequate) training of operators — four failures, including one prosecution and one prohibition.

(k)   Inadequate monitoring and/or testing of aqueous effluent before discharge, resulting in exceedance of process authorization and consent limits — three failures, including two prosecutions.

(l)   Failure of process or equipment design procedures (for example, at the design review stage) — three failures, including one prohibition.

(m)   Inadequacy of process record-keeping compared with process authorization requirements — two failures.

(n)   Discharge of aqueous effluent from an unauthorized (that is, not included in a process authorization) or unknown outfall — two failures.

(o)   Failure to provide operating staff with access to copies of process authorizations — two failures.

(p)   Failure of environment-critical equipment without back-up — two failures, including one prosecution.

(q) Operation of a prescribed process without a process authorization — one failure (prosecution).

INDIVIDUAL TYPES OF FAILURE: THE LESSONS

The failure types have been divided into three groups depending on the number of HMIP actions: large (that is, those which appear to occur most commonly), moderate and small. The division was approximately one half, one third and one sixth respectively of the total of 83 failures, and in the context of the incidents surveyed this indicates priority areas for attention. The failure types are indicated by (a), (b), etc as in the previous section.

The recommended actions are shown in italics.

MAJOR CATEGORIES OF FAILURE

This subsection considers the five most common failure types, covering 43 failures.

(a) Inadequate or non-existent operational procedures were a common HMIP complaint. A detailed examination of the press releases suggests that this problem arose more frequently in the peripheral areas of a process, rather than in the core processing area. Areas typically found to be lacking adequate procedures were:

- bulk liquid storage and handling;
- vehicle loading and offloading;
- effluent handling and treatment.

These operations should be reviewed and formal procedures prepared or updated with the same rigour as appropriate for core processes, and taking into account the potential environmental impact of the operations:

(i) Transfer operations within tankage areas (tank to tank), and between tankage and processes, particularly with respect to tank overfilling.

(ii) Vehicle loading and offloading (road, rail and ship), to ensure the proper connection of transfer lines (hoses, etc); the disposal where necessary of displaced vapour to destruction (for example, incinerator or flare) or recovery, or to a fixed vent for dispersion; the siting of vehicles such that accidental spillages will be contained.

(iii) Handling and disposal of accidental spillages, concentrating on their containment and recovery, or treatment prior to discharge, with minimum impact on effluent outfalls. This item refers also to spillages or emergency draining in process areas.

(iv) Operation of effluent handling and treatment systems to ensure optimum performance and retention of loads within the capabilities of the equipment.
Operating procedures are included in the Techniques of BATNEEC, and must be given the appropriate attention. While their development can be tedious and time-consuming, the detailed examination of procedures affords a means of turning over stones and dealing with any unpleasantness found underneath, before it can cause a problem for the process operator.

(b)   Failure to complete improvement programmes contained in authorizations appeared widely across the chemical manufacturing and waste disposal industries, and applied not only to process improvements but also to the provision of data on the environmental impact of processes. A number of precautions are necessary to avoid this type of failure, when improvement proposals are put forward by HMIP, and these can be built into an overall formal procedure for establishing or varying a process authorization.

(i)   Ensure that all improvements are included in a draft process authorization, including those put forward by the process operator when making the application, and that the completion date is that proposed by the applicant.

(ii)   Before accepting proposed improvements and their completion dates, discuss them in detail with operations, design and engineering divisions, and obtain their commitment to the programme. If this is not forthcoming, determine the reasons (for example, insufficient time for engineering completion, or lack of suitable plant shutdown timing) and present them, with alternative proposals, to HMIP. If necessary, request that the first phase of the improvement programme is the generation by the process operator, with a time limit, of an engineering programme for completion of the actual improvements required.

(iii)   If HMIP insists on including the programme that was originally proposed in the final authorization, consider an appeal against the unacceptable aspects of the programme, again with alternative proposals.

(iv)   If an initially acceptable programme shows clear signs of falling behind

schedule during execution, so that the completion date is unlikely to be met, discuss the underlying reasons for the delay with HMIP as soon as possible, with a new estimate for completion and a request for a Variation of the Authorization. It is essential to monitor the progress of improvement programmes carefully.

It must be recognized that the time limits applied to an improvement programme in a process authorization under IPC are binding, and failure to comply with them is an offence. A resultant Enforcement Notice may well include very stringent completion dates which will be difficult to meet.

(c) Emissions of vapour to atmosphere were mainly caused by the failure of abatement plant — for example, scrubbers or dust filters — or the absence of abatement plant complying with BATNEEC (EPA90) or BPM (HSW74).

The following actions should be taken by process operators:

(i) Ensure that, where gas or vapour are emitted to atmosphere, the abatement equipment provided at least meets the requirements of BATNEEC as described in the relevant HMIP process guidance notes. If it does not, then develop a plan to meet the emission standards, by installing new abatement plant or preferably by reducing the emission if possible. Discuss and agree the plan with HMIP.

(ii) Where abatement plant is installed, ensure that it is fully operational and not failing due to lack of proper maintenance. Put formal procedures in place for the routine inspection and maintenance of abatement plant. See also (d).

(iii) Ensure that instrumentation included in abatement plant is adequate to give warning of deteriorating performance or sudden failure, so that maintenance or process changes (for example, shutdown) can be implemented before a significant emission occurs. If suitable analytical instrumentation is not available, set up a formal sampling and testing schedule instead, and include it in plant operating procedures.

It is important to recognize that a process is authorized for operation on the basis that the plant is properly maintained, as well as correctly operated.

(d) Failure of maintenance procedures led to a number of emissions to air or water, either because on-line maintenance was inadequate, or because a process was restarted after a maintenance shutdown with abatement plant maintenance incomplete.

Process operators should:

(i) Put in place formal procedures for the routine on-line inspection and maintenance of abatement plant, including its instrumentation and control systems, to ensure that it continues to operate at high efficiency. These procedures should be integrated with those for the main process.

(ii) Insert in plant start-up procedures, formal checks that abatement plant maintenance is complete and that the plant is ready for operation. Abatement plant recommissioning should be an integral part of the overall process start-up procedure.

(iii) Ensure that stacks and vents for dispersion of emissions are internally cleaned regularly as part of normal maintenance procedures.

It is essential that abatement plant be treated as part of the overall process, and not as an add-on item of secondary importance.

(e) Uncontrolled, or inadequately controlled, discharges to water resulted from:
• a lack of facilities to retain spillages or emergency drainings, or spillages in areas where the drainage interceptor was designed to cope only with surface water;
• poor location and operation of tank bund drainage valves;
• discharge via settling and interceptor facilities which were not cleared of solids sufficiently frequently.

Process operators should:

(i) Ensure that areas where spillages can reasonably be expected are drained to a retention facility to allow recovery or controlled treatment of the spilled material. Alternatively, ensure that the effluent treatment system serving the areas is capable of taking a spill without loss of separation performance. Surface water interceptors generally are not designed to separate and retain large flows of chemical or hydrocarbon spills.

(ii) Check that tankage area bunds are provided with valved drains to effluent treatment, and that the valves are kept shut except during controlled draining of the bunded area. Draining routines should be included in tankage operating procedures. The valve positions (open or shut) should be easily observable, and the valves easily accessible.

(iii) Ensure that all effluent treatment systems are desludged regularly. Build-up of solids in separation and settling bays reduces the residence time of effluent in the bays, thus reducing separation efficiency and running the risk that outfall consent limits will be exceeded. Regular checking of sludge levels in bays should be included in operating procedures for effluent systems.

INTERMEDIATE CATEGORIES OF FAILURE

This subsection considers the next six fairly common failure types, covering 28 failures.

(f)   A process authorization usually includes a number of requirements for the correct operation of the process in order to minimize its environmental impact. These operating conditions are binding, and operating without one or more of them constitutes a breach of the conditions of authorization. Typical contraventions are:

• operating without temperature controls on bulk storage (for example, refrigeration of volatile materials);

• operating without temperature control on reactors (for example, operating reactor heating on manual instead of using the automatic controls provided);

• operating without abatement plant in good condition;

• ignoring restrictions on discharges to tidal waters.

It should also be realized that 'soft' requirements, such as the setting up of operating procedures by the process operator, are also binding.

Process operators should ensure that the binding nature of authorization operating conditions is recognized by their staff.

(g)   A common failure in observation of authorization operating conditions has been the inadequate measurement of emissions. Process authorizations contain requirements to measure and report emissions (concentrations or mass emissions) regularly (compliance reporting), using either instrumented or manual methods, which are prescribed in the authorization.

Where instrumentation fails or is found to be inadequate, an alternative method must be substituted — for example, manual sampling and laboratory testing. Regular measurement and reporting of emissions are ideal activities for control by a formal procedure and schedule, which should define the full analytical requirements (with analysis of water and oxygen in flue gas to allow correction to standard conditions), by whom and when the samples shall be taken and analysed, or readings taken, and the format and timing of reporting. The basic schedule is set down in the authorization, and must be complied with once it has been agreed via the draft authorization.

(h) A typical subset of failures relating to compliance reporting is late reporting of data. The time-base for reporting is binding, and must be complied with.

A procedure such as that described in (g) above ensures that reports are submitted on time.

(i) Compliance reporting is a regular activity, irrespective of whether the results are within the compliance limits or not. Where compliance limits are exceeded, however, or where an unauthorized emission occurs — for example, a spill to controlled water, or a serious smoke emission — the process operator is required by the authorization to report it in writing to HMIP within 24 hours of the operator becoming aware of the emission. This is a binding condition, and failure to comply can result in prosecution. The report may be made by fax, and should include a description of the incident and initial proposals to remedy the problem.

The reports are probably best made by a single focal point for the operating site. The authorization should also be checked to determine whether the same requirement applies to the National Rivers Authority for effluent failures and spillages, when the time limit may be very much shorter — for example, as soon as possible — to allow remediation where necessary.

(j) Inadequate training of operating staff was highlighted by HMIP, particularly with respect to procedures for dealing with spills or vapour emissions, and included staff other than process operators — for example, the competence of tanker drivers to deal with spills during loading and unloading was queried.

Process operators should ensure that they put in place adequate and suitable training for staff involved with operation of processes and peripheral activities, on both existing and new plants, and including environmental procedures. Formalize the training and keep records of levels of competence required and achieved.

(k) A number of cases were noted where effluent was discharged to controlled waters outside the conditions of the consent, due to inadequate testing or decision-making prior to discharge.

Where effluent is initially contained, a procedure should exist for testing the effluent to check if it is suitable for discharge. If it is not, the effluent must be treated to bring it into compliance and prevent further pollution. Where possible, continuous effluent flows should have analysers installed or a sampling and analysis routine set up, to give early warning of deteriorating effluent and the need for action, formalized in a procedure, to prevent a pollution incident.

MINOR CATEGORIES OF FAILURE

This subsection considers six less common failure types, covering 12 failures.

(l)   A small number of instances of process plant design failures were mentioned. Errors which could have been noted and eliminated at the design stage were not dealt with, leading to potential or actual emissions.

A design review should formally include a check for design faults which can lead directly to emissions, or can start a chain of control or operator actions which can lead indirectly to an emission. A check should also be made on the adequacy of the vent and drainage arrangements and the related abatement equipment.

(m)   Some authorizations require the logging (record-keeping) of environmentally-sensitive plant data — for example, the operating temperatures of an incinerator, or the circulation rate and strength of a vent scrubbing liquid. Such record-keeping is mandatory, and its scope will have been defined in the authorization.

The records should be retained as required and be made available to HMIP on demand.

(n)   There were two cases of effluent discharges to unknown or unexpected final discharge points, leading to pollution incidents. Ignorance is not a defence in these cases, and all discharge points must be declared when applying for a process authorization.

Before making an application, a process operator should ensure that drainage schemes are up to date and correct, and also survey the site and adjacent waterways for all outfalls, including land drains which may be draining what have now become potential spillage areas due to plant expansions. Keep drainage plans up to date, and make such surveys and updating part of the overall procedure for plant developments.

(o) There were two instances of copies of authorizations not being made available to process plant operators. Process authorizations require that direct operating staff should have access to copies in their work area. In one of the two instances, this had not happened 15 months after the issue of the authorization.

Ensure that copies are made available (for instance, in the shift foreman's office) and make them easily recognizable by using a specific colour of binder and a registration number. All copies should be part of a document control system, and no random copying should be allowed. When an authorization is updated, call all copies in and update them simultaneously.

(p) A lack of spares for critical equipment is asking for trouble.

At the design stage of a plant, give careful consideration to the necessity to provide full spares (installed or replacement parts) for all equipment which is environmentally critical, including instrumentation and control systems. Bear in mind the reliability of the critical equipment, and the impact of its failure — for example, plant shutdown — on the overall operation.

(q) Operation of a process plant without an authorization is almost certain to lead to prosecution .

It should be part of the initial design procedures of a plant to determine whether the new plant is an extension of an existing prescribed process, and hence requires a variation to an existing authorization, or whether it is a new prescribed process and thus requires a new authorization. In either case, open discussions with HMIP at an early stage to establish how to proceed.

CONCLUSIONS

This survey of the lessons to be learned from Prohibition Notices and other actions taken by HMIP shows that the Inspectorate places great emphasis on the establishment of formal operating procedures as a means of reducing the risk of environmental problems arising from the operation of a prescribed process. Detailed examination of the lessons indicates that many of the failures that occurred could have been avoided by the application of formal procedures to the design, operation and authorization aspects of processes, and that these procedures should extend over all aspects of the processes, including activities which are peripheral to the main processes — such as storage, transport and effluent treatment. In reality, all these peripheral activities should be treated as integral parts of the main processes, and receive the same emphasis.

Once these wide-ranging procedures have been established, they can be integrated into an overall environmental management system, capable of regular auditing and updating as process and regulatory changes occur. This gives the process operator increased protection against environmental incidents and legal enforcement of authority requirements.

ACKNOWLEDGEMENTS
The author wishes to express his appreciation of the help of the staff of the Press Office of HMIP for the prompt provision of the press releases used in this survey.

4. BS7750 AND CERTIFICATION — A DEVELOPMENTAL HISTORY

Christopher Sheldon

The British Standard specification for environmental management systems (EMS), known more commonly by its quality analogous number, 'BS7750', has had a fascinating parentage, a stormy adolescence and is now ready to go out into the world and seek its future. Chronicling some of that upbringing will not only help to understand what is happening now, but also what will happen beyond the publication of the international version of the standard, ISO14001. What has been learnt from this experience? How well are UK companies placed to tackle the requirements of BS7750? What are the problems that lie ahead?

Looking at the life cycle of BS7750 to this point, it falls neatly into two halves, both interdependent and yet both unique. The first is the inception of the standard, how it came to be written, and the way in which it functions within the 'multiverse' of standards, regulations and legislation. The second is the development of accredited certification, the seeking out of acceptable objective evidence that requirements have been met, and all the practical difficulties that have had to be overcome. The whole is a consistent, efficient tool that can be used by organizations who seek to improve their environmental management and performance on a continual basis.

THE BIRTH OF BS7750

The study of life cycle analysis has provided the phrase 'cradle to grave', and in this respect the cradle of BS7750 was a British Standards Institution (BSI) technical committee (originally designated EPC/50 within BSI and now referred to as ESS/1) formed in order to write the standard. As many may appreciate, the committee was put together using the procedures to ensure balanced representation, which is laid down for BSI in the standard for standards 'BS0'. Like all committees, there is only one BSI employee involved, who acts as the committee secretary, while the chairman is elected at the first meeting, choosing from those representatives who have been invited to join. The committee then decides its own working methods, writing capability and — to a certain extent — scope, with the secretary's role limited to facilitation and advice.

EPC/50 decided at its first meeting that the need for the standard was urgent enough to warrant the employment of outside consultants to produce a first draft document in response to a brief set out by the committee. This is well within normal work practices for BSI Standards committees, though under normal circumstances the drafting process is carried out by the committee itself. The expense of taking on such consultants, however, mitigates against their wider use. In this case, the committee felt the need for early production of a fully fledged standard that would give the UK a leading position, and so moved quickly.

The original brief to the drafting consultants makes interesting reading. It outlines the basic character of BS7750 and largely defines the nature of the document. The main points are set out here; it is interesting to note that, though the document itself went through several drafts, the brief has not only been met but is still current. The proposed EMS standard must:

- be compatible with ISO9000;
- be capable of stand-alone use;
- be applicable to manufacturing, process and service industries;
- consider the total organization;
- consider the total process;
- be capable of certification;
- be compatible with national and EC regulations.

GETTING THERE FIRST

The combined use of drafting consultants and the broad representation of industry, government and consumers on EPC/50 ensured that a standard that had gone through the full consultation process required was produced within twelve months of the first committee meeting. Published in April 1992, the standard was a world first, and in the next two years it was either adopted wholesale by other countries as a national standard (Holland, Denmark, Israel) or was the trigger for publication of a BS7750 clone standard (Ireland, France, South Africa). It is easy to forget that, at the time of publication, the International Organization for Standardization (ISO) had not even completed deliberations on its own work programme in the area of EMS and related business tools.

The advantages of being the first to publish in such an important and influential area cannot be over-emphasized, but it is also fair to say that there are some inherent disadvantages. One of the problems was that the standard had not been thoroughly tested by companies on the ground. This is why the original standard was published with a special preface that stated BS7750 would be reviewed and possibly revised in the light of user experience within twelve months. This allowed a year of industrial implementation experience to be fed back into the standard writing procedure to ensure that any practical problems were ironed out.

PILOT IMPLEMENTATION SCHEME

In order to keep a clear route for the great mass of reports from companies that would be coming to EPC/50, BSI Standards arranged a special facilitation process, which rapidly became known as the BS7750 Pilot Implementation Scheme. An open invitation was issued to any interested companies and organizations to gather together into working groups, elect their own chairmen, and report to a steering committee that would ensure consistency of approach in the final reports, as well as helping sort out any practical queries during the process of trial implementation.

Each working group was put together using the 'Nomenclature générale des activitiés economiques dans les Communautés Européennes' (NACE) sector headings as laid out in the (then) draft Eco-Management and Audit Scheme regulation (EMAS), to ensure consistency with that document's approach. Obviously, as the working groups were not put together under the aegis of BS0, they could not be considered representative of a particular industry sector in the manner of a standards committee. But it did mean that industry had an organized channel of information, into which experiences could be directly fed. Over 450 organizations spread over 37 different sectors of industrial activity took part, from April 1992 until April 1993.

When the final meetings took place under the Pilot Implementation Scheme in 1993, the exercise had certainly fulfilled its earlier intent. All 37 working groups submitted a report to EPC/50 with comments on their experience; they suggested changes to the standard and further comments on the role of any intended certification. The comments could be addressed in one of three ways:

• by changing the wording of the standard;
• by the writing of an industry sector owned 'Sector Application Guide';
• during the certification process.

The technical committee's role meant that it could only take cognisance of the first category, as its remit was defined by the ability to change the wording of the standard. If the answer to the problem area identified lay outside this remit, then there was little the committee could do. This meant that there was a large amount of work on developing other parts of a self-regulatory system to support the standard that still remained to be done even after the revised standard was published in January 1994.

It is fair to say that many companies who took part in the Pilot Implementation Scheme had unrealistic expectations that they would be certified to the new standard by the end of the scheme. Several aspects of the work mitigated against this. The first and most important was that, during the twelve months of the Pilot Implementation Scheme, several pieces of both national and European legislation and regulation were either being drafted or implemented.

Firstly, there was the developing European Regulation concerning EMAS which, although heavily influenced by BS7750 (indeed, before the standard was published the regulation covered only auditing procedures, and scarcely mentioned EMS, let alone specified a standardized management system), was still open to further change. It was a core element of the original brief from EPC/50 that the standard would be compatible with all European legislation, which meant a careful examination of the development of the document through to its final publication in June 1993.

At the same time, at a national level, Her Majesty's Inspectorate of Pollution (HMIP) was developing the implementation of integrated pollution control as instigated by the Environmental Protection Act 1990. This too had elements that called for continual improvement, and though HMIP was kept closely informed of the development of BS7750, there was still, as already noted, a lack of practical implementation of the self-regulatory framework of certification. There was also much development work still to be done on the interpretation of the regulatory framework (such as the continued publication of Chief Inspector's Guidance Notes).

Internationally, by the time BS7750 was published in its revised form, ISO had embarked on an ambitious programme of standards writing in the area of environmental business tools — the proposed ISO14000 series. The international equivalent of an EMS standard (ISO14001) was well under way, and though the European contingent was looking for a standard to support EMAS, American and Japanese delegations scented a possible barrier to trade being erected, and began to question the whole basis of such a standard. Had they been successful in their efforts, it is likely that Europe would have had to produce its own European standard, or at the very least an annex, in order to support its own

Regulation. This would have made a mockery of the whole standardization process. In fact, the resulting draft international standard is largely compatible with both EMAS and BS7750, but back in January 1994 the situation was not quite so clear cut.

MOVING TOWARDS CERTIFICATION

The exact placing of the new standard in relation to national, international and European initiatives therefore depended largely on the development of independent third party certification to the standard itself. Implementation of the standard in such a way that companies would be able to supply objective evidence of their ability to meet its requirements would mean meeting many practical problems head on, especially where consistency of interpretation was required. Within BSI, the baton of developmental responsibility was passed from BSI Standards to an internal office responsible for environmental service development. The Corporate Environment Office, headed by the Environment Policy Manager, was responsible for the next phase of development within BSI, which led to BSI's accredited certification service available today.

The fact that certification was thought necessary right from the inception of the standard is worth noting. Again, referring to EPC/50's original drafting brief, the value of the standard was closely identified with its ability to be used within a certification context. Industrial self-regulation demands increasing amounts of credibility, especially in an area where competing 'green' claims are still a matter of concern for the Department of the Environment and the Department of Trade and Industry alike. Were certification left undeveloped, BS7750 could simply have become a simple aid to self-declaration, as indeed is allowed by the current draft of ISO14001. However, the difference between 'self-declared' and 'unsubstantiated' to clients and customers in the modern market-place is virtually non-existent.

All this indicated that BS7750's credibility could only be supported by a certification system, allowing independent third parties to assess organizations against the standard and issue certificates to those who complied with the requirements. This in turn required the development of a national accreditation system to set standards of performance for the certification bodies themselves. Many have commented on the fact that BS7750 has key phrases that appear at first sight to defy objective definition — phrases such as 'significant effects'. Certification thus presents an opportunity for purely subjective assessment, and it became of paramount importance to ensure consistent practice across the spectrum of certification activity.

It was with this in mind that BSI resisted pressure to begin certification activities immediately on publication of the revised standard. Instead it decided that it would not issue certificates until a national accreditation scheme existed, such as the one supported and administered by the National Accreditation Council for Certification Bodies (NACCB) for ISO9000. It came as no surprise, therefore, when the Department of Trade and Industry, after issuing a consultation document, chose to extend the remit of the NACCB to include such an accreditation scheme, adding to it the responsibility of accrediting individuals and organizations who wished to provide verification services in support of EMAS.

In June 1994 the NACCB called together all the organizations interested in developing and offering certification services and invited them to take part in their own 'pilot phase', during which accreditation criteria would be developed. The organizations were mostly those bodies already involved in certification to ISO9000, with an additional smattering of environmental consultants eager to set up certification businesses. Draft accreditation criteria were published in November 1994, following a series of trial assessments of companies by the certification bodies witnessed by NACCB. The finalized criteria were published in January 1995, after the field trials had been supported by a series of certification body 'head office' assessment visits, again carried out by NACCB assessors. Following recommendations made by the assessors to NACCB's Environmental Accreditation Panel, the first wave of organizations were presented with their accreditation certificates on 8 March 1995, and certification began in earnest.

MAKING CERTIFICATION WORK

Both NACCB and BSI realized from the start of this cycle of development that the focus of the work had to be ensuring that assessment and registration against BS7750 was more than an empty, paper-driven exercise. There was no point in issuing certificates if the environmental performance of the company in question simply failed to improve. The problem was one of emphasizing and seeking out signs of continual improvement against the company's own stated environmental policy in as objective a manner as possible without demanding anything that was not already a requirement of the standard. The image of a company retaining its BS7750 certificate for year-on-year improvement of its office paper recycling haunted all concerned. The watchwords became consistency and credibility.

The developmental work was obviously extremely valuable in turning the basic theory enshrined in the pages of the standard into hard practice through the certification process. What BSI learnt in the process of making certification a real and viable proposition brought tangible benefits to both its clients in particular and the practice of environmental management in general. An overview of this experience may help to understand more closely the nature of EMS certification.

BSI undertook some 17 trial assessments as part of the trial phase of development, across a wide spectrum of industrial activities, including companies in the chemical sector. This not only gave great insight into the nature of those seeking registration to BS7750, but enabled a certification process to be devised that not only met the requirements of the standard, but the nature of the clients and the readiness (or otherwise) of their management systems.

The process (shown in diagrammatic form in Figure 4.1 on page 48) acknowledges that the system under scrutiny is not only very much broader in coverage than, say, an equivalent quality management system, but also that its development will in all probability be at a much earlier stage. In other words, companies confronting BS7750 for the first time — even those with considerable experience with ISO9000 — are relative novices when it comes to applying the principles of an EMS. As this is also true of BSI's own certification process, extra lateral movement in the form of a multi-stage process (allowing for a greater number of possible deferments called for either by BSI or its clients) takes into account the need for a gradual learning process for all concerned.

MULTI-PHASE ASSESSMENT

Three phases of assessment can be run as three separate visits (currently recommended to first-time clients) or can be combined within a two-visit structure, providing there is ample evidence in advance that the system and its implementation are at a suitably developed stage. After each phase, depending on what has been discovered, a deferment can be called for by either the assessment team or the client. This allows non-conformances with the standard identified within that phase to be corrected and followed through by the client. The length of each deferment will obviously be related to the nature and extent of the non-conformances discovered, as well as the ability of the company to undertake the necessary remedial activity.

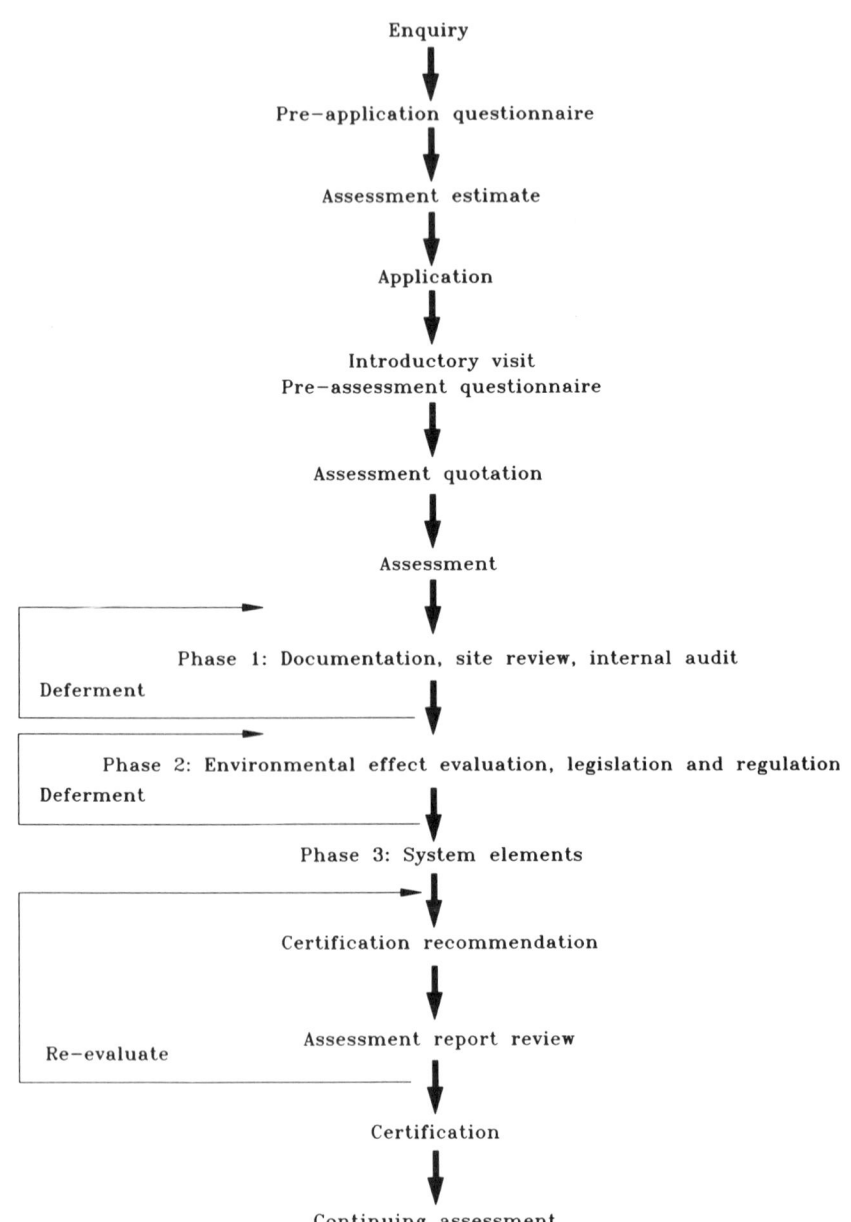

Figure 4.1   BSI QA's route to registration.

This multi-stage process was developed following the experience of sending in teams for a single visit assessment after an ISO9000 style 'pre-assessment' visit, only to discover that the earlier visit had failed to discover basic flaws in parts of the management system under scrutiny. The scope of an ISO9000 'pre-assessment' was obviously not sufficient to flush out, say, a problem in identifying and evaluating environmental effects as required by Clause 4.4.2 in the standard. And there was more. Although the ISO9000 pre-assessment visit involves a tour of the site, the level of detail required by a site tour for BS7750 is much more intensive and requires far more time to undertake thoroughly. The company's own internal audits of an EMS proved crucial in forming an estimate of how long the final assessment might take — the more effective the audit, the more time could be saved by the assessment team refraining from duplicating the entire exercise.

PHASE ONE

The first phase of BSI's newly-designed BS7750 process requires assessors to undertake documentary reviews while actually on site, at the same time giving careful scrutiny to the client's own internal audits, and an in-depth tour of the whole site. This gives both BSI and the company a much better picture of the overall scope of the management system, allows for large gaps in the system to be addressed at an early stage, and sets the scene for the next stage of the assessment. It also means that certain issues such as the boundaries of the site can be agreed (not always easy if there are other leased buildings within the company's site), as well as ensuring that no 'ring fencing' of activities has taken place upon the site.

Some of these interpretive areas are still open to the decision of the individual certification body, but many of the major areas of concern are covered by the requirements of the NACCB, which has, for instance, set an upper limit on the amount of time (20% of the total) that can be deducted from an assessment as a direct result of a company's internal EMS audit. The NACCB has also undertaken to review its accreditation criteria and policy guidance within the next eighteen months in the light of further certification experience.

PHASE TWO

The second phase of certification is probably the most important to first-time users of the standard. It addresses the core elements of the EMS and how it relates to the legislative and regulatory framework applicable to its scope. BSI discovered that many companies had focused very tightly on the production of a register of significant environmental effects, sometimes to the detriment of the procedures and criteria that supported its compilation. Many simply did not realize that the key part of the requirement is the establishment and maintenance of procedures and criteria to identify and evaluate all the organization's environmental effects, and that the compilation of the register should spring from that, not the other way round. It could be said that the requirement for a written register is the tip of the iceberg, providing external assessors with the objective evidence required to prove that a particular process has taken place. Without the register, assessors do not have a coherent focus on which to base their examination of the procedures that resulted in the compilation of the company's own environmental effects.

In any case, the examination of the register of significant effects and the subsequent acceptance of the register by the visiting assessors helps to define the parameters of the management system under scrutiny. If there is a significant flaw in the register, or in the procedures that support it, there is little point in assessors continuing any further, as it is a vital part of the core EMS. Management systems are rather like computers: 'garbage in, garbage out'. If a company has misidentified or underestimated the extent and range of its environmental effects, then the system designed to manage the effects will share similar blind spots and inadequacies. A fault in this element of the system is almost always considered a major non-conformity.

Many of the points concerning the register of significant effects apply also to the register of legislation. Again, the register is a requirement of the standard that supplies an external assessor with objective evidence that the company is aware of all environmental legislation and regulation that applies to it, and that it has the procedures to ensure that it keeps abreast of legal developments. The exact scope of the legislation — as it relates to specific industrial activities and as it is defined in documentation — isn't always clear, and can itself be open to interpretation.

A good example is the way in which best practicable environmental option (BPEO) is an inherent requirement of the way in which integrated pollution control is enacted, based on the Environmental Protection Act 1990. BPEO is certainly further defined in very helpful Chief Inspector's Guidance Notes issued by HMIP, but simple possession of the appropriate notes says little or nothing about a company's ability or intent to comply with such guidance. In other words, a principle that can only truly be realized through implementation cannot adequately be assessed against the mere holding of a document. A register of legislation does mean, however, that a company cannot claim ignorance as a defence if a subsequent implementational flaw relating to legal requirements within the management system is discovered at a later stage. Obviously it depends very much on the circumstances of the company concerned, but the baseline requirement of BS7750 is an adherence to the law. The scope of the register thus needs careful examination and agreement between certification body and client before assessment can proceed.

PHASE THREE

The final phase of assessment then turns to what one could euphemistically call the 'mechanical delivery' of the management programme and policy defined by the effects evaluation and legislative scope. Once the first two phases of certification have been gone through and the basic parameters identified and agreed, this part of the assessment largely covers the same elements as an ISO9000 assessment. It would be wrong to think, however, that assessors involved in this stage are interchangeable with ISO9000 assessors, as the objective evidence being sought is of a very different nature.

It is fair to say that, as the collective experience of both BSI and UK industry grows, the nature of the certification process will change. At this early stage there is still much to be discussed and learnt from practical implementation of the standard against assessable criteria. As with quality management, where change has occurred over the last decade, there is no doubt that further experience will dictate changes to such criteria. It is in this area that sector application guides can be developed as the most effective disseminators of best practice within a defined industry.

The original BS7750 Pilot Implementation Scheme, though now well and truly over, provided the basis for some of the work in the further development of these documents. Individual working groups agreed to carry on meeting on their own initiative to develop guidance documentation independent of BSI Standards. Until now, however, these guides have been operating with only half the story. Advice on implementation prior to the finalization of the certification process was always going to be, at best, inspired guesswork. That is not to say that the guides published so far have not had their function, though the quality of the advice on offer has been extremely variable. Sector application guides should aim to be, essentially, short shelf-life documents. They should bring those using BS7750 up to a level of knowledge that allows them to apply the standard in such a way that the requirements for objective assessment are satisfied. They are more closely linked to the certification process than the standard, and should be used in that context.

WHAT HAPPENS NEXT?

Now that certification has begun in earnest, what benefits will it bring? Perhaps the biggest benefit will be the eighteen month lead in implementation that the UK now has on the rest of the world. The current draft of ISO14001 and interest in BS7750 from both Japan and America indicates that the rest of the world is moving towards the EMS model that the UK and BSI have been examining and developing intensively for the past three years.

At the level of the individual company, the use of BS7750 will provide a framework for detailed and exacting cost-benefit analysis of environmental management. Beyond the much-touted benefits of a systematic approach to such cost-saving areas as energy efficiency and waste minimization, continual improvement will require the efficient costing of new processes, new plant and the associated risk assessment. The social expectation of a leading company may centre on its environmental performance, but the management of costs is left to managers themselves. Continued business survival now depends on adaptability and efficiency as never before, with the extra dimension of environmental expectation. The longer a company resists considering environmental aspects of its operations, the costlier the penalties for late entry will become. They will take the form of tighter legislation, rising customer expectation and more expensive end-of-pipe solutions to basic process problems.

There will, of course, be those that wait for the level playing field, arguing that continual improvement is an endless escalator on to which they are not prepared to step. This is an argument for inaction. In the UK, continual improvement is already an integral part of our national legislation. In Europe, leaders in industrial efficiency such as Germany and Holland are getting ready to take EMS and EMAS on board as soon as possible. Even more significantly, a number of large Japanese corporations are already piloting the use and certification of BS7750 in their UK plants with a view to taking it across Europe and eventually the world. It appears, once again, that global industrial leaders do not wait until the camber of the playing field is to their liking.

If this sounds like an advert for a UK success story, it is. EMS is an area where the UK leads the rest of the world, and the future for those with the vision to embrace it looks particularly bright. The developmental history of BS7750 and attendant certification certainly bears out the words of Theodore Hesburgh, 'The very essence of leadership is that you have to have vision. You can't blow an uncertain trumpet'.

5. THE DEVELOPMENT OF INTERNATIONAL STANDARDS

David Hunt and Catherine Johnson

This chapter summarizes recent developments in the standardization of environmental management systems and related areas, both in the UK and internationally. Since BS7750 was first published early in 1992, related developments elsewhere have proceeded apace. Several other countries — including Canada, France, Ireland and South Africa — have been working concurrently on their own national environmental management systems (EMS) standards. The International Organization for Standardization (ISO) has been working on various aspects of environmental management, firstly through its Strategic Advisory Group on the Environment (SAGE), and subsequently through its Technical Committee TC207 on Environmental Management.

Within Europe, the European standards body, Comité Européen de Normalisation (CEN) is following the work of ISO to see whether or not its own efforts will be required to produce separate regional standards to underpin the EC Eco-Management and Audit (EMA) Regulation.

Subsequent sections of this chapter describe the current state of these various initiatives and activities, and consider likely future developments.

ISO ACTIVITIES

ISO established Technical Committee TC207 in June 1993. This committee, for which Canada provides the secretariat, has six Sub-Committees (SCs) and a Working Group of its own, as listed in Table 5.1 on page 56. The work of SCs 1, 2 and 4, addressing environmental management systems and closely related issues, is described below.

ISO TC207 SC1 — ENVIRONMENTAL MANAGEMENT SYSTEMS

The work of SC1 has been undertaken by two Working Groups. Working Group 1 is producing a Specification which prescribes only those elements of an EMS which may be audited, together with guidance on the use of the Specification. Working Group 2 is producing more general guidance, to include details of good EMS practice, and the facilitation of management culture change.

TABLE 5.1
Sub-Committees of TC207 and their Working Groups

Sub-Committee and Secretariat	Subject	Working Groups of the Sub-Committees*
SC1 — UK	Environmental management systems	1: Specification (and guidance on its use) 2: General guidance
SC2 — Netherlands	Environmental auditing	1: General principles 2: Audit procedures 3: Qualification criteria for auditors
SC3 — Australia	Environmental labelling	1: Guiding principles (programmes and systems) 2: Type II labelling 3: Basic principles of all environmental labelling
SC4 — USA	Environmental performance evaluation	1: Generic performance evaluation 2: Industry sector performance evaluation
SC5 — France	Life cycle assessment	1: General principles and procedures 2: Inventory (general) 3: Inventory (specific to manufacturing operations) 4: Impact assessment 5: Life cycle improvement assessment
SC6 — Norway	Terms and definitions	None

* There is also a Working Group of TC207 itself, addressing environmental aspects in product standards, for which Germany provides the secretariat.

SAGE bequeathed to Working Group 1 a 'model for discussion' on similar lines to BS7750. Working Group 1 considered this, together with national EMS standards and drafts. The outcome — after several stages — was a draft Specification (ISO/CD14001), considered further on page 59.

Similarly, Working Group 2 produced a draft of its document on general guidance (ISO/CD14000). Both Working Groups were asked to consider the particular needs of small- and medium-sized organizations, but there will not necessarily be a separate standard or guideline on this subject.

At present, the Committee drafts (of both ISO14000 and ISO14001) are undergoing a period of public consultation. National views will be considered by SC1 at its next meeting in June 1995, after which a further Committee draft or a Draft International Standard could be issued. The current draft of ISO14001 is discussed later (see page 59), in relation to its compatibility with BS7750 and the EC Eco-Management and Audit (EMA) Regulation.

ISO TC207 SC2 — ENVIRONMENTAL AUDITING (EA)

SC2 has established three Working Groups. The first is addressing General Principles, the second Audit Procedures (of EMSs, in the first instance — see below) and the third Qualification Criteria for Environmental Auditors. The documents being produced by these Working Groups are summarized in Table 5.2. A fourth Working Group has been established to consider audit-related activities, such as site investigations.

The first three Working Groups have each produced one Committee Draft, comments upon which have been examined. New Committee Drafts, produced in October 1994, have recently been approved by SC2 and are now

TABLE 5.2
Environmental auditing standards currently planned by
ISO TC207 Sub-Committee 2

Type	Working Group	Proposed number	Subject
Generic	1	ISO14010	General principles of environmental auditing
Procedures	2	ISO14011/1	Environmental management system audits
Auditor qualifications	3	ISO14012	Qualification criteria for environmental auditors
Audit-related activities	4	Unknown	Activities such as site investigations

undergoing a further period of public comment. This includes an invitation to comment on the need for separate standards on General Principles and EMS audits, as opposed to a single standard if the detailed procedure for EMS audits is sufficiently generic to cover all types of environmental audit. The need for additional work on Compliance Audits and Environmental Statement Audits is to be considered by Working Group 2; Working Group 1 will consider the need for documentation on Initial Reviews and Working Group 4 will similarly address the need for documentation on Site Assessment.

All these auditing standards are planned to be Guidelines rather than Specifications, an intention confirmed by the comments received on the Committee Drafts. The comments also confirmed that, whilst supporting the EMS standards to be produced by SC1, the auditing standards should be capable of use with a wider range of EMS standards and models. ·

ISO TC207 SC4 — ENVIRONMENTAL PERFORMANCE EVALUATION
Environmental performance evaluation (EPE) is the process of measuring, analysing, assessing and describing an organization's environmental performance against agreed criteria for appropriate management purposes. The work of this SC therefore has particularly important links with the work of SC1 on EMS and SC2 on EA, and the goal is to provide assistance to organizations in the design and implementation of their own performance evaluation.

Initial work addressed such issues as:
- purposes for which EPE is undertaken;
- areas/categories in which performance indicators may be needed;
- general types of performance indicator which may be used;
- selection of performance indicators.

A framework document was produced, but further work has been hampered by the lack of any pre-existing documents with a consensus backing to form the basis for further work. Nevertheless, a draft document taking account of the 100 or so documents and comments received so far was produced before the meeting of SC4 in March 1995.

It had been proposed to proceed to development of performance indicators, both general and sector-specific, but this is a controversial issue with opinions varying widely from country to country. The SC aims to produce a Committee Draft on environmental performance evaluation by early 1996.

DEVELOPMENTS RELATED TO THE EMA REGULATION

The EC Eco-Management and Audit (EMA) Regulation was published in July 1993 and became operational in April 1995. Its Article 12 on 'Relationship with national, European and international standards' provides for organizations achieving accredited certification to EMS and EA standards to be considered as meeting the corresponding requirements of the Regulation, *provided* that the standards are approved by the Commission.

One of the main reasons for the early revision of BS7750, beginning in 1993, was to ensure its continuing compatibility with the EMA Regulation, which was only in draft form when the Standard was first published, but which was finalized prior to publication of the 1994 version.

It is understood that the 'Article 19 Committee' (responsible for setting up the measures needed to bring the Regulation into effect) has agreed that BS7750 meets the corresponding requirements of the EMA Regulation, and that this decision will shortly be formally announced. This will end a period of uncertainty and reflect BSI work to ensure that the Standard is compatible with the Regulation.

The EMA Regulation mandates CEN to produce an EMS standard to support operation of the EMA Scheme. In view of the resources required to produce a European standard, and the desirability of a global EMS standard if it can be achieved within ISO, CEN is reluctant to start any new work unless the need for a separate European standard is clear. At the time of writing, CEN has been debating the compatibility of ISO14001 and EMA, and is now awaiting an opinion from the 'Article 19 Committee' set up to determine the acceptability or otherwise of particular EMS standards in relation to EMA requirements.

COMPATIBILITY OF BS7750, ISO/CD14001 AND EMA

Within TC207 SC1 Working Group 1 there has been considerable debate concerning, in particular, the degree of detail to be incorporated in an ISO EMS Specification. Efforts have been made to resolve significant differences of approach through detailed discussions of individual requirements, as all participants recognize the value of an agreed worldwide EMS Specification if such can be achieved.

There are subtle differences in wording between ISO14001 and BS7750 in a number of areas, but the fundamental EMS model is essentially the same in both documents, and in the EMA Regulation. Some areas of difference include the following:

CONTINUAL IMPROVEMENT

BS7750 and EMA explicitly require an organization to make a commitment to continual improvement in environmental performance. More specifically, EMA and BS7750 refer to reducing adverse environmental effects to ' ... levels not exceeding those corresponding to economically viable application of best available technology' (or EVABAT for short).

As one might expect, national or regional concepts such as BATNEEC and EVABAT have not found their way into ISO/CD14001, which defines continual improvement in terms of enhancing the EMS with the purpose of achieving improvements in overall environmental performance.

DEFINITION OF 'ORGANIZATION'

EMA defines the organization to which it applies in terms of a site; BS7750 allows for a multi-site organization. ISO/CD14001 defines an organization so that this may be a single 'operating unit', even if this occupies only part of a site, formally recognizing what may have to be acknowledged in practice in the implementation of both BS7750 and EMA.

ENVIRONMENTAL EFFECTS EVALUATION

ISO/CD14001 has adopted the word 'impacts', with essentially the same meaning as 'effects' in BS7750, and a new term 'aspects' referring to components of a company's activities, products and services which are likely to interact with the environment.

BS7750 specifically requires procedures for identifying, examining and evaluating all effects — both direct and indirect — but has in its 1994 revision added the guidance that the evaluation of indirect effects should include all those which it can control, or could reasonably be expected to influence. The EMA Regulation does not so specifically refer to indirect effects in general, though its Annex D of 'good management practices' refers to 'any significant impact on the environment in general' and to advice to customers on the handling, use and disposal of products.

The current ISO approach is subtly different from both. It requires identification of 'aspects' of the organization's activities, products and services that it can control or influence (test of reasonableness included), and then a determination of those which have (or could have) significant impacts (that is, effects). This might be considered a more logical approach than those of BS7750 or EMA, avoiding the need to consider effects over which the organization has no control or influence, but it is to be hoped that it does not discourage broad examination of effects.

Furthermore, ISO/CD14001 appears to make less specific mention of contractors (in BS7750 and EMA) or suppliers (in BS7750), and of covering the consequences of past, as well as of current and future, activities (covered explicitly by both BS7750 and EMA) — although these aspects are not excluded by the definition of 'environmental impact'.

REGISTRATION OF ENVIRONMENTAL EFFECTS/IMPACTS
BS7750 and EMA require the organization to maintain an internal register of significant environmental effects. The ISO approach to date calls for a procedure to identify environmental aspects to determine which have significant impacts; it does not, however, call for them to be recorded.

REGISTRATION OF LEGISLATIVE AND REGULATORY REQUIREMENTS
BS7750 and EMA require a record to be kept of all applicable environmental legislation and regulations. ISO/CD14001 requires a procedure to 'identify and have access to' such material, but not to record it.

TIME-SCALED OBJECTIVES
BS7750 and EMA refer to objectives quantifying (wherever practicable) the commitment to continual improvement over defined time-scales. ISO/CD14001 retains the quantification requirement (with the practicability qualification), but refers to time frames in relation to the environmental management programme devised to meet the objectives.

PUBLIC AVAILABILITY OF OBJECTIVES
BS7750 requires that objectives are publicly available and, to facilitate this, the revision of 1994 also added that policy statements should tell the reader from where the objectives can be obtained. EMA appears likewise to require that objectives be made publicly available. This is not stated explicitly, but is implied by a combination of elements in Articles 1, 3 and 5.

ISO/CD14001 shares with BS7750 the requirement to establish, maintain and document objectives (and targets); currently, however, it does not appear that it requires objectives to be made publicly available. This seems likely to be a drawback to those whom BS7750 defines as 'interested parties'. It is at odds with the increasing trend towards environmental performance reporting;

the number of companies which are publishing their own environmental performance record, often against their own publicly declared objectives, is growing all the time.

MONITORING AND MEASUREMENT
All three documents (BS7750, EMA, ISO/CD14001) differ in this area. BS7750 is most detailed, referring to (amongst other things) specification of requirements, monitoring procedures and quality control procedures. The EMA Regulation refers to requirements and monitoring procedures, and ISO/CD14001 refers to establishing monitoring procedures and calibrating equipment — which is only a part of measurement quality control.

TIMETABLES FOR IMPLEMENTATION
In ISO terminology, ISO14000 and 14001 are 'Committee Drafts' and the current 'Committee Stage' continues until comments are considered, a consensus achieved and a draft International Standard is put forward. Under ISO rules, the Committee Stage may last anything up to seven years! In practice it is likely to take much less, but even after publication as a draft International Standard the text still has to be approved by the responsible Technical Committee, and this stage — including commenting and voting — takes time. Nevertheless, whilst it is now nearly three years since BS7750 was first published, an international convergence of all the environmental management initiatives now under way could, in principle, be achieved by late 1996.

Table 5.3 shows the main timetables involved in the UK. Consultation on an accreditation system for BS7750 took place in 1993, and it has been known for some time that the National Accreditation Council for Certification Bodies (NACCB) will operate this system. A six-month pilot certification programme began in May 1994, involving about 20 bodies intending to offer certification to the standard.

For EMA, events are proceeding broadly in parallel; systems for the accreditation and supervision of verifiers were due to be in place by April 1995, when the scheme began to operate.

At the international level, the earliest that ISO TC207 SC1 could produce a specification on EMS is about two years after agreement of a Committee Draft — probably mid-1996 at the earliest, subject to a consensus being reached on the draft in early 1995. Whether or not a separate European (CEN) EMS

TABLE 5.3

Timetables for UK implementation of various initiatives

Event	1992	1993	1994	1995	1996
BS7750 published ... and revised	●		●		
Consultation on accreditation		●●●			
EMA Regulation published		●			
BS7750 pilot certification programme			●●●		
BS7750 accreditation operational?				●	
EMA becomes operational				●	
ISO EA standard published?					●
ISO EMS standard published?					●

standard is needed will depend upon the extent to which an eventual ISO standard is considered likely to meet the requirements of EMA, as discussed above.

ACCREDITATION ARRANGEMENTS

The NACCB has been designated as the accreditation body for BS7750, and a pilot certification programme was conducted in 1994. Following this, the proposed accreditation criteria were published, for comment, in November 1994. These specify in some detail the systems and competencies required of certifying bodies and the duties placed upon them. They also describe how compliance or otherwise of a client's EMS BS7750 or EMA is to be determined in certain instances.

63

The proposed criteria draw particular attention to legislation and regulations applying to the environment within the workplace, including COSHH, Management of Health and Safety at Work Regulations 1992 and the Workplace Regulations. This seems to imply that certifying bodies will need to have the capability to assess health and safety, as well as environmental, management systems. BS7750 specifically states in its Foreword that, 'This standard is not primarily intended to address, and does not include requirements for aspects of, occupational health and safety management; however, neither does it seek to prevent an organization from incorporating such issues into its environmental management system'. However, a guide to occupational health and safety management systems, based on the ISO9000 approach, has been drafted for public comment by BSI.

JOINT HEALTH, SAFETY AND ENVIRONMENTAL MANAGEMENT SYSTEMS

In this context, it is interesting to note that several industries have produced guidance on joint health, safety and environmental management systems (the Chemical Industry's Responsible Care programme, and the upstream oil and gas industry's HSEMS Guidelines, produced by the Exploration and Production Forum).

This raises the question, 'Can we expect an ISO Standard providing a generic Management System (MS) specification, with associated sub-specifications or guidelines for the application of such a specification to quality, environment, health and safety and (perhaps) other matters?'. The best answer that can currently be given is, 'Possibly, but not for a long time'.

The ISO Technical Committees responsible for quality management and environmental management, TC176 and TC207, are liaising. The issues involved are complex, however, and the progress of international standards development is inevitably relatively slow. These factors seem likely to ensure that the production of such a generic 'umbrella' MS standard, with suitable revision of the existing quality management standards and draft environmental management standards, is many years away. Nevertheless, the way forward is pointed by the October 1994 production of a draft ISO Standard specifically for the offshore petroleum and natural gas industries, dealing jointly with health, safety and the environment, based on the HSEMS Guidelines of the Exploration and Production Forum, cited above.

CONCLUSIONS

There is now widespread recognition of the importance of the EMS in enabling companies and other organizations to manage their environmental affairs proactively and effectively. There is also a substantial measure of international agreement in all areas of environmental management — and rapid progress, in the context of standards making. Only about three years after the publication of the first national standard, BS7750, there is vigorous and productive international activity across a wide range of related issues.

Whilst the availability of certification and verification to recognized patterns is important, the fundamental role of standards in providing guidance and models to follow should not be forgotten. There are, as many companies have shown, tangible business benefits to be achieved from sound environmental management, in addition to those provided by registration under a formal scheme.

6. SURVEY OF INDUSTRIAL EXPERIENCE WITH ENVIRONMENTAL MANAGEMENT

Cynthia Riemann and Paul Sharratt

This chapter details the results of a survey which was conducted to obtain an overview of the ways in which companies are currently addressing environmental performance and management. Fifteen companies were surveyed with the intention of capturing a 'snapshot' of the current position with regard to environmental management, viewing which alternatives are being considered and pursued, and assessing the success of the results that have been achieved with the options chosen.

The results of the survey were compiled with the aim of providing companies with a benchmark for environmental efforts, identifying any trends that might be emerging in environmental management in the process industries, defining some of the key arguments about environmental management systems, and showing ideas and alternatives that might be used in developing a system of environmental management.

INTRODUCTION

With the increase of pressure on industry to achieve and demonstrate sound environmental performance and with the publication of such environmental management systems standards as BS7750, the European Union Eco-Management and Audit Scheme and ISO14000, environmental management has become a central issue in many companies in the process industries. Although environmental management is not a new issue, many companies are now concentrating a great deal of effort on addressing the questions of how to manage and control the environmental impacts of their operations and how to handle environmental issues.

There is considerable variety of opinion and experience regarding what factors are to be considered and what decisions are to be made in approaching this task. There is also considerable debate over how environmental control and management can best be achieved. Many solutions have been offered: some companies are implementing independent, formal environmental management systems based on the BS7750 model; other companies are extending their safety and health management systems to include environmental considerations; and still others are deciding that no formal system is needed at all. It was with the idea of taking an overview of this variety in industry that this survey was conducted.

SURVEY STRUCTURE

Fifteen companies in the process industries were examined as part of the survey. The questions that were posed to the companies centred around the goals of:
• defining the status of environmental management within the company;
• examining general considerations and systems decisions made about environmental management;
• determining company commitment and resources given to this task;
• identifying strategies employed in managing environmental concerns.

Although a standard question-set was established for the interviews, there was a great variety of situations and experiences. Subsequently, the interviews generally took on a less constrained conversational form rather than a strict question/answer style.

The survey was conducted by site interviews (though in a few cases, when site interviews were impossible to arrange, phone interviews were substituted). Those interviewed within the companies were environmental managers, environmental personnel, and managers who could provide an accurate picture of the company's environmental management efforts. This method of survey was chosen in preference to a more general questionnaire format in order to obtain detailed and specific information regarding environmental management.

Rather than focus on one particular industry sector, the companies surveyed were chosen from throughout the process industries to give a broad cross-section of experience with environmental management. Figure 6.1 shows the operations of the companies examined. (For reasons of confidentiality, it was not possible to be more specific about the companies interviewed.) Of the fifteen companies surveyed, six were exclusively UK companies and nine were multinational companies. The size of their businesses ranged from single-site plants to massive, global operations. This set was not intended to be statistically representative of British industry, but rather an overview of industrial experience with environmental management from which generalizations might be drawn.

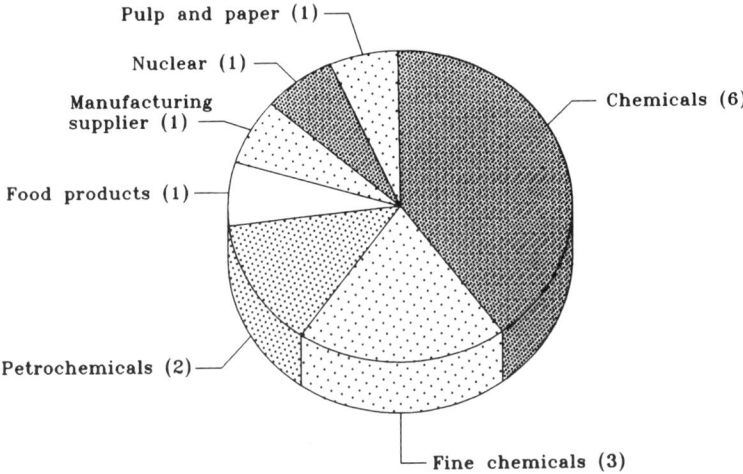

Figure 6.1   Industries surveyed (numbers in parenthesis are the number of companies in the category).

GENERAL FINDINGS

The most immediate, though not surprising, finding of the interviews was a confirmation of the extraordinary differences from organization to organization in commitment, approach, decisions and progress made with regard to environmental management and control. The responses from companies on environmental issues covered a spectrum – from those who expressed a very low, secondary commitment to those who had a very high degree of corporate environmental commitment and where environmental concerns were given a high profile within the company. Actions taken by companies ranged from 'staying one step ahead of the law' to allocating extensive resources for environmental concerns and implementing formal systems of environmental management.

Despite the considerable variety of approaches to environmental management, some general categories of response could be distinguished. Figure 6.2 on page 70 illustrates the actions of the companies. The various categories of response can be described as follows:

Implementing (2)
These companies found that a formal, independent, accredited environmental

management system was desirable and was in line with the organization's operations. They were in the process of implementing BS7750 into their operations.

Considering (4)
These companies also found that a formal, independent, accredited environmental management system was desirable and was in line with the organization's operations. At the time of interview, they were either deciding between several systems or were watching the progress of the 'Implementing' category to survey their experience.

Safety/Health/Environment (5)
These companies saw the need for a formal system of environmental management, but they believed that environmental concerns would best be addressed as part of a combined safety, health and environment system.

No formal system (3)
These companies did not see a need for a formal system of environmental management within the company. Although they did have some control procedures for environmental issues, they did not sense any pressure to develop a complete framework for an environmental system.

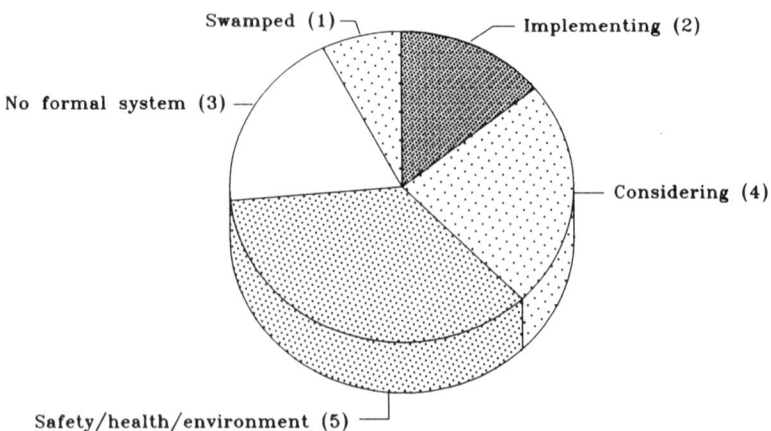

Figure 6.2   Systems summary (numbers in parenthesis are the number of companies in the category).

Swamped (1)
This company was so inundated with other business concerns that it barely had time to consider even basic environmental issues. Its strategy was one of 'fire-fighting' only.

SYSTEMS AND SPECIFICS
The remaining survey results can be divided into two broad categories:
• findings relating to overall system considerations and structure; and
• findings relating to specific points of environmental management.
    The first category looks at the variety of approaches in choosing or not choosing a certain form of system to aid in environmental management. The second examines how companies are addressing specific issues common to most systems of environmental management. These two categories are addressed separately in the remainder of this chapter.

SYSTEM CONSIDERATIONS
The various system choices of the companies with regard to environmental management highlighted an important division in industry over where and how to place environmental management within an organization. The question addressed was one of deciding if a formal system of environmental management was desirable; and if it was, then how such a system should operate both on its own and within the overall organization to achieve the company's environmental objectives. From the responses to the survey, the system choices focused on a structural management approach. There were three general approaches considered by the companies interviewed: establishing independent environmental systems, developing an integrated system or using no formal system. Figure 6.3 (see page 72) shows the further partition of the interviewed companies into structural approaches to environmental management.

INDEPENDENT SYSTEMS
Those companies establishing environmental management systems such as BS7750 ('Implementing' and 'Considering' categories) were putting into place a system that would operate in a fairly stand-alone fashion. This system, like others within the organization, would have primarily independent operating procedures and control mechanisms and would not be intertwined with other

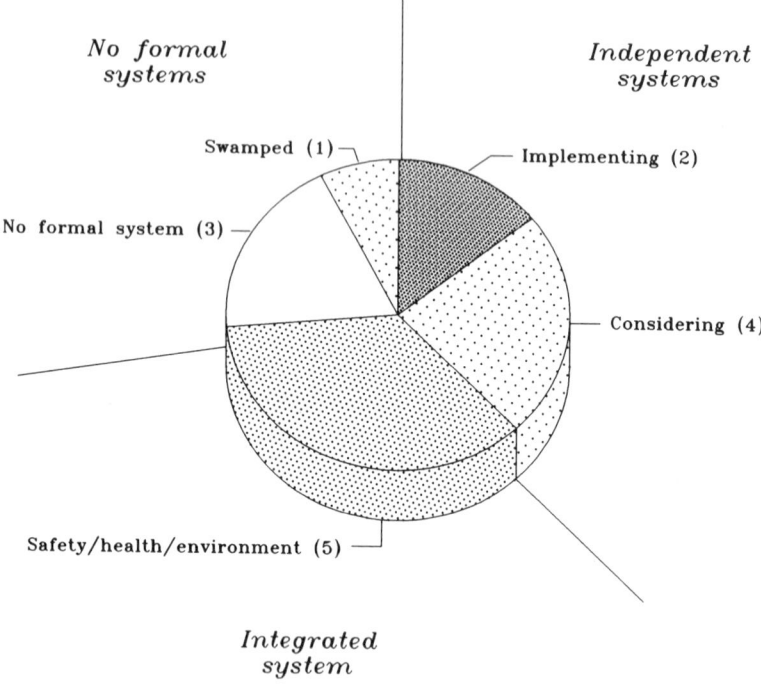

No formal systems

Independent systems

Swamped (1)

Implementing (2)

No formal system (3)

Considering (4)

Safety/health/environment (5)

Integrated system

Figure 6.3   Structural divisions (numbers in parenthesis are the number of companies in the category).

management systems and functions (such as safety, health or quality). The environmental management system, however, would not exist in complete isolation from other company operations. It would still be consistent with general company procedures and if co-ordination or information exchange were necessary for control of environmental issues (or conversely, if environmental information were necessary for another management control function), it would occur. Ultimately, though, the environmental management system would be autonomous within the organization.

INTEGRATED SYSTEM
Those companies developing an integrated system ('Safety/Health/Environment' category) saw no need for a separate system for environmental management. What they desired was a single system of management that would provide

all of the necessary procedures and control mechanisms for directing matters that required management within the company. Under this system, safety, health, environmental and other issues would be considered and managed collectively. Overall management practice would be determined for general control of operations and then necessary provisions for the management of environmental issues would be placed within that system. The majority of companies with integrated systems saw an independent environmental management system as a waste of company resources and an unnecessary duplication of practices.

NO FORMAL SYSTEMS

Those companies with no formal environmental system (independent, integrated or otherwise) did not see a need for a complete system to manage environmental concerns. This does not imply that they had a complete lack of organization or that there was an absence of any co-ordinated control efforts for environmental concerns; it means that they did not sense any pressure to develop a complete framework for an environmental system. Their view was that if they could keep up with legislation and permitting requirements without a system, then that was enough.

Further consideration of structural divisions is given in the following sections.

INDEPENDENT SYSTEMS

Given the current attention toward environmental management systems from the publication of standards like BS7750, many of the survey questions centred around finding out the various advantages and disadvantages claimed for using such a system. This section investigates further the reasons why companies chose a 'stand-alone' system; it presents the benefits listed by the companies for such a system, presents the reservations that some companies have in considering an environmental management system and lists the drawbacks given for such a system.

REASONS

Numerous reasons were provided by companies for their choice of pursuing an independent environmental management system to be used alongside other systems. The responses involved both company business practices and environmental pressures. First, many stated that a more independent system would fit well with their existing operations and systems; it was a natural extension of what was already being followed within the company. Separate departments already existed for these functions so it followed that an 'independent' systems approach was the simplest and most consistent with the organization for them. Second, many companies felt that a separate, accredited system was the way in which public and corporate opinion and favour was moving; many envisaged potential customer pressure for certification of such a system. Third, several noted that given the separate arrangement of the various regulatory systems under which their industries operated, it was useful to keep their systems separate in order to reduce confusion and focus their efforts. Finally, to several the idea of a separate system made organizational sense; to paraphrase one manager, 'It is useful to start with a discreet focus to ensure thoroughness. You then may be able to combine practices later, once you see how they might work together'.

BENEFITS

Both of the companies which were implementing environmental management systems had chosen BS7750. Of the benefits listed by these companies (and by those considering implementing systems) for an accredited environmental management system such as BS7750, those most frequently cited were competitiveness and marketing advantages, although environmental improvement issues were mentioned. The benefits included:

- good public and customer relations — most companies felt that they needed 'to be seen to be green'. Implementing an environmental management system was seen as a means of giving proof to their customers and to the public of environmental commitment;
- competitive advantage of being certified — since these companies felt that an accredited system was the way in which public and corporate favour might be moving, they decided to ensure that they could continue to do business by implementing a system for which they could be certified;
- competitive advantage of being ahead on the environmental curve — the general attitude with regard to environmental performance was that those who are in the environmental lead are less likely to be 'caught out' with stricter competitive environmental demands;
- good framework for pursuing environmental improvements — these systems were seen by those implementing them as ultimately able to facilitate long-term environmental performance improvement and benefits;

• going to do it anyway — in some cases, it was felt that the measures required within a certified system were tasks that would need to be done in the company anyway; therefore, it would be advantageous to receive certification for just doing what was already necessary.

RESERVATIONS

Those companies which favoured an independent environmental management system, but were still considering the specifics of such systems before proceeding with implementation, had several reservations. These reservations involved corporate environmental systems, different opinions of international operators and certification questions. All of the companies in the 'Considering' category were part of large organizations. As part of their operations, most of the companies had a formal, corporately-written environmental management system with which they had to comply. These companies had made the decision to consider implementing an accredited environmental management system in addition to their corporate system. Quite understandably, one basis on which the companies were screening possible environmental systems was the ease with which the corporate system could be adapted to achieve certification. Another basis for comparison was the pressures from various international operations. For example, some companies' European operations favoured the European EMA scheme, while their UK operations preferred BS7750. Although certification under BS7750 would qualify under EMA, they felt that the different systems have some significantly distinct provisions. (One company considered the BS7750 environmental effects register to be too extensive, but also found the EMA provisions for public environmental statement to be too demanding.) Those with American counterparts heard strong complaints about any system that would require listing of impacts that was not legislatively mandated (the idea being 'never to admit to any impacts'). They heard even more protest over the idea of a public environmental statement.

Additionally, the 'Considering' companies had no desire to be at the 'cutting edge' of environmental systems, preferring to see what happened with those who were first applying for certification. They hoped to see how certification would generally proceed and to learn what problems occurred for the 'first-round' companies so, if need be, they could react accordingly. Finally, several companies were waiting until customer pressure forced them to put in place a system capable of achieving certification.

DRAWBACKS

The 'Considering' companies were not the only ones that had some reservations about environmental management systems. Each company interviewed listed drawbacks to the systems, including those which were currently implementing them. One of the most frequent doubts expressed was that the cost of putting such a system in place would not be justified by improvement in environmental performance, by environmental cost savings or by increased competitive advantage. Additionally, many companies questioned the meaning of certification from a system like BS7750 which does not establish minimums of environmental performance or expected improvement. (It was an argument also associated with the BS5750 quality assurance standard — that certification guarantees little to the customer and does little good for the companies' operations.) Lastly, there were many concerns expressed over the objectivity and uniformity of the certification process. A number of companies interviewed speculated that it would be an extremely difficult task to make an objective decision on certification given the vast difference in business operations and the amount of documentation to be checked.

The majority of drawbacks listed by the companies centred around resources allocation and funding. Many companies contended that a BS7750-style system would require undue resources to implement and to maintain, likening it to a 'personnel vacuum' and 'paper mountain'. Ultimately, the issue was one of cost. Many companies thought that they could not spare the money to implement a system that, for the most part, was not fundamental to their business; in the words of one manager, 'Although we accept environmental responsibility for our operations, we are in the business to make money, not to be charitable to the environment'. The cost estimated (and reported) for implementing an environmental management system ran from tens to hundreds of thousands of pounds.

INTEGRATED SYSTEM

One alternative proposed to a stand-alone system was a completely integrated system of management. Such a system would provide operating procedures and control mechanisms to direct the issues within the company requiring management, including environmental issues. The companies choosing such an integrated system of management had decided upon what functions needed to be performed within the organizations (planning, organizing, implementing and controlling functions). They had created an overall system of management procedures for those operations and then incorporated aspects of environmental performance requiring management within that system. The system covered safety and health issues as well.

The advantages listed for such a system were:

• uniformity of procedures — with one general set of overall procedures, there are no significant differences in practice from one area in the company to another;

• simplified control — organization control is more easily achieved if all operations are handled together 'under one roof';

• efficient use of resources — there is no duplication of resources to manage various aspects of company operations. There is not the cost of maintaining several systems with several sets of procedures;

• stream-lined operations — with integration of management procedures, lessons from one field can be readily passed on to another field (for example, safety lessons can be used in developing environmental controls), keeping operations stream-lined;

• grouping of similar concerns — it is easier to group similar management functions for different elements together given an integrated system.

An interesting common point between several of the companies choosing an integrated approach was that they had high-hazard operations. Accordingly, most had long-standing, well-developed safety systems in place within their organization. To them, the environmental concerns of their operations were closely linked to hazard concerns; therefore, it made sense to add environmental considerations to their safety systems, rather than create an independent system for environmental management. (The exception to this was one small company which was developing a system for safety and environmental concerns at the same time and decided that it would be best to consolidate the areas into one system.)

SPECIFIC POINTS OF ENVIRONMENTAL MANAGEMENT

Although not all companies followed the strict provisions of an environmental management system such as BS7750, for the sake of analysis several specific points of environmental management were chosen from around the BS7750 'loop' for examination. These points were chosen as representing some possible features of environmental management, regardless of structural approach. They were particularly chosen to highlight the different practices and alternatives in environmental management. The areas selected were policy, objectives and targets, environmental effects register, environmental effects evaluation, design/management of change, customer/supplier environmental performance, training and auditing.

POLICY

All companies interviewed had something that could be pointed to as an environmental policy. (Even the company categorized as 'Swamped' had committed something to paper.) The policies varied from vague 'one-liners' of intentions of environmental benevolence to fairly specific statements, signed by the highest levels of management, which included explicit environmental 'action items'.

OBJECTIVES AND TARGETS

The task of establishing environmental objectives and setting environmental targets was labelled as 'easy' to 'incredibly difficult'. A number of survey participants found it convenient to elevate items reported to the Chemical Industries Association as part of their Responsible Care Programme to the status of company environmental objectives and targets. Not surprisingly, most environmental targets focused on easily-measurable quantities like money and/or tonnes, with several companies expressing concern over the difficulty of setting environmental targets that were not easily quantifiable (for example, it is easy to measure the number of hours of environmental training, but not the effectiveness of that training.) Lastly, one sceptical manager frankly commented that environmental objectives and targets were only going to be set in line with what the company had already planned to do, or with what would be required in legislation or permitting — just phrased more enthusiastically.

      The idea of continual improvement was agreed with in principle; however, several companies, due to their particular business constraints, found the idea somewhat impractical for their business. Industries operating in relatively 'clean' businesses, companies operating under best available technology and industries that were unable to change their process due to registration considerations found that continual improvement to the environmental performance of their operations was difficult, if not impossible to achieve. Often, environmental improvement does not follow a linear path — for example, from high emissions directly to zero emissions — but levels out due to process limitations; further improvement is not achievable without a change in technology. This would result in a step change in performance, not in steady improvement over time. This step change could then be followed by improvement only to the point that it too levels out due to the limitations of the process. It was this idea that was of concern to many businesses.

ENVIRONMENTAL EFFECTS REGISTER

There was a general uneasiness among those interviewed as to what would constitute 'enough' with regard to registering and recording environmental effects. The questions heard included:

- what effects should be included?
- what is significant?
- how far into past activities (and how far forward into future effects) should we go?
- what about suppliers' activities and effects?
- how specific should we be with indirect effects?
- what constitutes an 'indirect effect'?

Most companies generally believed that they had been given very little practical guidance in this matter and that this task would be extremely difficult in practice.

ENVIRONMENTAL EFFECTS EVALUATION

The evaluation of environmental effects drew, like no other, a resounding cry of 'Help!' from those interviewed. Uniformly, companies conveyed that they were having great difficulty assessing the environmental impacts of their operations and products. They felt that there was very little practical guidance available (there was general dissatisfaction with the proposed best practicable environmental option (BPEO) index), that the entire area was too new and ill defined and that means of evaluation were not very strongly based in scientific quantification of effect. Many believed that complete evaluation of their environmental effects was too much to expect, given the techniques and assistance resources available. (One manager expressed amazement that HMIP and operators were given the task of evaluating environmental impacts at all, when so few techniques and so little information are available.)

DESIGN/MANAGEMENT OF CHANGE

A number of companies had considered the environmental impact of a new process at the design stage (and some, to a degree, at process changes.) These considerations extended as far as establishing a rating system within the design process that took into account environmental impacts and considered only process routes which were environmentally 'more friendly' in process determination and in product formulation.

CUSTOMERS/SUPPLIER PERFORMANCE

Most of the companies interviewed had been questioned about their environmental performance by customers. (Those who had not been asked questions were generally those who supplied their products internally to another part of their business.) Most questions posed to the companies were found not to be very demanding (for example, yes/no inquiries such as 'Do you have an environmental policy?' or 'Do you have an environmental program?'). However, one company was questioned with regard to one of its processes and was asked to change the process to lessen the overall impact. Luckily, the buying company was willing to absorb the cost of the extra charges incurred in production. (It might be of interest to note that this was a chemical product that was being produced for a major company.)

Some companies questioned the environmental performance of their suppliers; most admitted, however, that it was not a very strenuous questioning. Several companies had issued 'warnings' to suppliers regarding their environmental performance, but only one company had made any significant decisions based on supplier response.

TRAINING

Almost all of the companies interviewed had fairly extensive safety training; many were looking to extend this policy to environmental training as well. But the majority of companies had very little environmental training for personnel at the time of interview. Most efforts involved training operators in the use and importance of environmental monitoring equipment, though some environmental awareness training was being given to senior-level staff. The general exception to this was that many companies had sent personnel for external specialist training in various environmental areas of company concern.

AUDITING

Environmental auditing was an area of considerable variation among those surveyed. All of the companies interviewed had conducted some form of environmental audit or assessment of their operations, with many reporting that they performed regular, periodic environmental audits. The complexity and thoroughness of the audits ranged from quick overviews of 'problem areas' to highly detailed accounting of environmental impacts. Audits were performed by company personnel, by consultants, and by using numerical scoring systems. There was no consensus about which approach had worked best.

CONCLUSIONS

Given the great variety of industrial experience, progress and understanding with regard to environmental management, it was difficult to draw many firm generalizations from the survey results. But that variety itself makes an important point about how environmental issues are being addressed within companies in the process industries. There is no consensus in industry about what is required for good environmental management, let alone what methods or systems might be best for achieving it. There appear to be several different, effective and equally valid approaches to managing and improving environmental performance that go beyond the provisions of any one system. Given the diversity of operations, corporate management cultures and resource bases, it seems unlikely that any single system of environmental management will be a good fit for everyone. Indeed, it might potentially be harmful to attempt to impose one model of environmental management onto all companies, regardless of their operations.

One definite conclusion did, however, come from the interviews: the need for better development of techniques for evaluation of environmental impact. Uniformly, those interviewed expressed a great need for more complete scientific information about environmental impacts, for better evaluation techniques and for more consistent procedures and expectations of what an evaluation of environmental impact can provide. Without this kind of support, many companies fear that environmental impact evaluation may be dominated more by speculation than by facts.

7. MAKING ENVIRONMENTAL MANAGEMENT WORK

Stuart Page

Nationally and internationally, the emphasis in environmental protection is moving steadily away from just achieving compliance with statutory release limits. It is now being placed on industry taking the initiative to prevent pollution in the first place, and to conserve and enhance resources. For industry, implementing a formalized, systematic approach to environmental management is a means by which the objectives of pollution prevention and sustainable growth may be achieved. This has been widely recognized, resulting in a drive towards standard specifications for environmental management systems such as BS7750, the Eco-Management and Audit scheme (EMA) and ISO14001. The incentives for company participation in such schemes are not limited to improved environmental performance. Benefits to business operating costs may also accrue as a result of, for instance, reduced wastage and increased efficiency. Additionally, certification to a recognized standard is intended to provide product market advantage to participating companies.

The concept of adopting a systematic approach to environmental management is an entirely appropriate means for business to achieve improved environmental performance. The standard specifications for environmental management systems essentially represent common sense and best management practice. For industry, however, the logistics of implementing a formalized system present a very significant obstacle to realizing the ideal, and the use of fully integrated systems is still far from widespread.

This chapter aims to highlight some important practical considerations for the effective implementation of environmental management systems, drawing primarily on experience within Courtaulds, the chemicals and speciality materials manufacturer.

ENVIRONMENTAL MANAGEMENT IN COURTAULDS

Courtaulds is a large organization with sites worldwide manufacturing coatings, fibres, films and a range of other performance chemicals and materials. The matrix of operating businesses within the corporate body is served by a number of centrally-co-ordinated facilities, including Courtaulds Health, Environment and Safety Services (CHESS). CHESS represents both a resource of environmental expertise for businesses and also a co-ordinator of environmental strategy across the Group.

A primary component of the Group's strategic direction in managing environmental issues is the introduction of more systematic and formalized management methods into all of the Group's businesses. CHESS has formulated an internal specification for an environmental management system, based on good management practice and incorporating the now familiar common elements of BS7750, EMA and ISO standards — that is:

- policy and objectives;
- resources;
- operating procedure structure;
- measurement and understanding of effects;
- improvement programmes;
- documentation;
- audit system;
- training;
- communication.

The aim in specifying this system is to produce a general standard which is applicable to any of Courtaulds' operations worldwide, is as straightforward to implement as possible and — with some further refinements — will enable the businesses to achieve certification to the available standards.

As already stated, however, the crucial aspect of operating an environmental management system is not in the theory or specification, which essentially represents common sense, but in the 'nuts and bolts' of establishing the system in a real business and making it work. In this respect, two operations within Courtaulds represent useful case study material — The Amtico Company Ltd, and Courtaulds Fibers Inc. These operations are significantly different from each other in a number of respects.

THE AMTICO COMPANY LTD

Amtico manufactures high quality floor tiles. Located in Coventry, UK, Amtico has a work force of around 300, and participated in the BS7750 pilot scheme. Amtico is aiming to achieve certification to the BS7750 standard in the summer of 1995. This operation is the furthest advanced business within Courtaulds in formalizing an integrated approach to environmental management.

COURTAULDS FIBERS INC (CFI)

This business operates a large viscose rayon fibre manufacturing site in Mobile, USA. Production involves complex and large-scale chemical processes, and the site qualifies as a Major Source under the US Clean Air Act Amendments. The Mobile site is in the early stages of establishing a systematic approach to environmental management; some elements of the system are currently well developed, others less so.

PRACTICAL ELEMENTS OF AN ENVIRONMENTAL MANAGEMENT SYSTEM

FOCUS AND COMMITMENT

The whole system starts with commitment at the most senior business level. The commitment must be based on senior management recognizing business needs, which may become apparent as a result of regulatory pressures, through an appreciation of potential savings, or through customer requirements for some form of product certification.

For many businesses, the recognition that there is a need in the first place may be some way off. The day-to-day business focus in these situations is most likely to be on productivity, quality and the more immediate cost issues of running a business. It is here that the advantage of a business being a part of a larger, more aware organization becomes apparent. The larger organization may be a corporate body, but could equally well be a trade or industry association.

There is a strong belief centrally within Courtaulds that the incentive for implementing an environmental management system should not primarily be in order to gain third party certification. Rather, the real benefits which are achievable at the bottom line of a business should provide this incentive. As an example, at CFI in Mobile the increasing complexity and stringency of environmental regulatory requirements has encouraged the business to look for novel and integrated solutions to environmental control. This in turn has required a more flexible and systematic approach to environmental management, specifically in process design. Consequently, a solution which allows the objective of reducing releases of carbon disulphide to air by 70% to be achieved will also enable major improvements in raw materials usages and process efficiency.

For some businesses, the prospect of some form of product certification may be particularly attractive, particularly in the case of products with a very strong brand identity. In other situations, the requirement for certification may be specified by customers. The leverage that certification will ultimately apply to encourage participation in, for instance, BS7750 is yet to be established. On the basis of experience with ISO9001, however, the potential is considerable.

CURRENT POSITION AND SCOPE

Identifying needs in terms of managing environmental issues requires some understanding of the current position. A pre-commitment review highlights the key components of a systematic approach to environmental management which are in place in the current operation. It indicates the capability of the current system and gives an appreciation of the scope of the task in hand. This latter aspect is particularly important; even if management are satisfied that operating an environmental management system will benefit the company, there is no point in proceeding if at the outset the resource to make it work is unavailable. This pre-commitment review may not need to be performed as a discrete exercise; management may well have this information already.

It is important at this stage to recognize the concept of stakeholders. Previously, environmental policy within many organizations was focused on achieving legislative release limits, and essentially placating the public. A systematic approach to environmental management needs to consider the effects of the operation as perceived by the public, employees, regulators, local residents, customers and financial institutions.

So who should perform this task? A small but experienced team of fairly senior operations and environmental managers would appear to be appropriate, with additional input from the corporate central environmental resource or from external consultants.

For large organizations incorporating a number of businesses, policy statements are often very general in order to be widely applicable. The business policy statement should recognize the scope of the undertaking and, where derived from corporate policy, should be specific enough to reflect the particular activities of the business in question.

Based on the policy, the initial set of objectives should be set to reflect the management perception of scope, in order that real progress can be made and demonstrated. The initial objectives should therefore be set on the basis of both environmental priority and the potential for making real, demonstrable progress. Progress which is demonstrated at an early stage acts as encouragement to all those participating in the programme.

DEVELOPING THE SYSTEM

Within the existing management structure, some functions which represent key components of any system of environmental management will already be in place. Individuals within the organization will have responsibility for environmental control, development, reporting and so on, and some network of internal reporting may also be in place. Whatever structure is finally adopted for environmental management, it is sensible to build on these existing components in order to minimize the size of the task, and to reduce the disruption to existing management responsibilities. The particular hierarchical structure will therefore be determined to a large degree by existing management arrangements at the site in question.

At CFI in Mobile, the aim is to build on existing line management responsibilities by formally drawing together an environmental responsibility matrix specifying who does what, where, when and why. This establishes both responsibility and accountability. The intention is to reinforce this flat structure with team-orientated task groups.

In initially developing the management system, the elements which already exist must be identified. There are then two options:

(1) to build on these elements; or

(2) to fill in the missing elements and 'complete the cycle'.

For many businesses, these missing elements will most likely be:

(i)   a procedure for understanding and recording environmental effects;

(ii) a procedure for understanding and recording regulations;

(iii) a formalized auditing procedure.

A systematic approach to management giving continual improvement cannot be established without these elements in place. It is therefore preferable to 'complete the cycle' of environmental management by establishing and linking together the elements specified above. No matter that this may well result initially in a complete system containing elements which perform less than optimally; if policy and objectives are set, responsibilities defined, and audits conducted, then the whole process will develop positively in an iterative manner.

A major consideration is whether to develop a management system that covers quality, health, safety and environmental issues together, or keep these separate? At the outset, it would appear to be most practical to keep the initial development of a system for environmental management separate from other management systems. Such an approach provides focus and defines the scope of the task to facilitate progress to be made. In most situations, it will probably be necessary to start by treating environmental management as a new initiative, but with the long-term goal of integration. Ultimately, however, this may not be practical. A fully integrated health, safety, environment and quality management system may exceed the critical mass for effective operation (the 'dilute and disperse' principle where management effort becomes spread too thinly across too many areas).

As to which individuals constitute the environmental management hierarchy, this has already been addressed in part. Clearly, specific individuals should be designated as responsible for certain environmental issues on the basis of existing responsibilities and expertise. In these cases, environmental objectives should be written into the individuals' job descriptions, and their performance reviewed and appraised on this basis. Nevertheless, environmental awareness and minimizing environmental impact should form a part of everyone's job. This can be facilitated through regular and appropriate communication and through training. Encouraging a high level of environmental awareness in its widest sense will help to commit the workforce to improved environmental performance.

At Amtico in Coventry, an environmental advisory panel has been established, chaired by an environmental management representative who is also a board director. The panel ensures that environmental policy is implemented and that line management and employees are supported in achieving environmental objectives. The panel is composed of a number of individuals covering a broad range of administrative, technical, managerial and commercial disciplines. A key member of the panel is a 'green group' representative; an employee who has a keen interest in environmental matters. The green group representative provides an alternative perspective on environmental matters, and acts as a focal point for informal communication between management and employees.

BUILDING INFORMATION AND UNDERSTANDING: CONTINUAL REVIEW
OF REGULATIONS AND EFFECTS
Understanding the environmental effects of an operation in as wide a sense as possible is vital so that objectives can be prioritized to enable the impact of the operation to be reduced. For most industrial operations, however, any attempt to quantify such effects fully would prove to be a truly Herculean task. In most cases, the scope must therefore be limited to identifying and ranking the significant major effects.

The effects to be considered include all issues of relevance to the stakeholders. In the UK, integrated pollution control (IPC) requires prescribed processes to render harmless any releases which cannot be prevented in the first place. In order to render them harmless, it is clear that the effects of the releases must be understood. IPC sites should therefore understand the effects of their releases on the environment, and certainly the public will want to know about them. The public and other stakeholders will also be interested, however, in noise, odours, use of non-renewable resources, waste, etc, and these will all have to be addressed in the assessment.

In order that an appropriate and complete list of effects be compiled, expert input is required. For Courtaulds businesses, this is available either within the business itself or from CHESS. Nevertheless, the opinion of the 'person in the street' should never be discounted. Employees are essentially 'people in the street' outside their work place, and so it may prove useful to canvas employee opinions on what they believe to be the important environmental effects of the operation. Again, this has an added benefit of raising the profile of the initiative and encouraging a wide range of involvement.

The major problem in assessing effects is not in identification, but in assessing significance. This requires some degree of quantitative comparison of effects. Methodologies have been, and continue to be, developed to quantify at least in a comparative manner the environmental effects of operations. The protocol proposed by Her Majesty's Inspectorate of Pollution (HMIP) for assessing the best practicable environmental option (BPEO) for process release, and life cycle analysis, are examples. In both of these cases, however, the usefulness of the technique is compromised by a paucity of good, reliable existing data.

Ultimately, therefore, the assessment of effects must be performed partly on a qualitative and subjective basis. In order to assess the overall impact of the site on the environment, Amtico have adopted a four-stage approach:

(1) identify goods, processes, buildings and operations on the site and in the immediate vicinity (1 km radius) which have effects;

(2) determine which of these are significant by considering whether abnormal conditions related to each would require either the use of the emergency services, informing the regulatory authorities, stimulate media interest, or cause an intrusive effect on the local ecosystem;

(3) assign a rating for each significant effect (1 for slight effects, 5 for extreme);

(4) average the data to give an overall site rating.

Clearly, this approach is in no way absolute, but it does provide a record of the rationale used to identify significant effects, and the reasoning behind some of the specific environmental performance objectives which have been targeted by the company.

Once targets have been established, an effective system of monitoring must be put in place to demonstrate that the targets are being met. Merely ensuring that measurements are taken will effectively bring the issues of control and improvement to the attention of those with responsibilities in this area. It is largely true to say that 'what gets measured gets done'; measurement is one of the most important elements in any approach to achieving continual improvement.

Maintaining a register of the environmental legislation of relevance to an operation — and more importantly maintaining an awareness of the requirements of the legislation at all levels in the company — should just be a matter of assigning specific responsibility and ensuring efficient communication. In practice, however, a first review of the environmental regulations of relevance to an operation will almost always produce some surprises. Where the surprises are of the non-compliance variety, needless to say that the necessary corrective action should go to the top of the objectives list!

COMMUNICATION

Senior management commitment provides the spark for an integrated approach to environmental management. 'Completing the cycle' by establishing and linking the key elements of systematic management gets the system working. However, extensive and appropriate communication is a pre-requisite to ensure that the system works well.

Often in industry, initiatives have failed because employees who represent vital components of the system have not been informed of what is happening, why, what they are required to do, and how things are progressing. It is ironic that this is often the easiest issue to address, and yet the most neglected.

Communication can be made to be effective on a number of levels: by incorporating relevant issues into work objectives; by disseminating information at regular team reviews; by regular bulletin-board updates; through newsletters; and via annual business reports.

All of these means are used within Courtaulds, some initiatives being site- or business-based, others used across the whole corporate body. For example, CHESS issues a bi-monthly newsletter, *Environment Matters*, which reports on Group-wide environmental issues. Further coverage of environmental issues is given in site and Group news sheets.

Sites are now developing environmental awareness training programmes, where individuals with specific expertise make presentations to employees on a range of relevant issues. In addition to improving communication and raising awareness, this has encouraged local, informal environmental task groups to be established within businesses.

OPERATIONAL CONTROL AND DOCUMENTATION

There is considerable overlap between the requirements for environmental and quality management in respect of operational control and documentation. All operating procedures should be written down and must be appropriate, complete, correct and verified to be so. They should be regularly reviewed, and clearly disseminated to those employees having responsibility for operation. Auditing will in part allow for this verification, but a separate system should be established to perform this duty more rigorously and to review and update procedures.

Adequate training must be made available to ensure high standards of operational control. This includes regular refresher training to maintain these standards. Another important way of maintaining standards is for managers to 'walk the talk'; management visibly demonstrating to the work force that environmental issues are high priority is a powerful means of gaining commitment. The flip-side to this is that management actions and decisions which are seen to be in conflict with policy or objectives will greatly weaken the overall commitment to continual improvement.

Responsibilities must be defined clearly to ensure that high standards of record-keeping and documentation are maintained. A quantitative approach — for example, statistical process control — in many instances facilitates high standards of both control and documentation.

The development of an environmental manual for the operation should be undertaken with careful reference to existing quality and safety manuals and documentation to avoid duplication. The final product needs to be user-friendly, widely available throughout the organization and relevant if it is to have anything other than ornamental use.

In order to be appropriate for their purposes and to continue to be relevant, all these elements of control and documentation must be checked routinely and frequently as part of the audit system.

AUDITS

Conducting audits and acting on the findings of audits is key to environmental management. Managers must be prepared to find deficiencies, and to act to correct these. This is a difficult nettle for some sites to grasp, however, where the perception (and sometimes the reality) is that audit findings may leave the business vulnerable to litigation. Nevertheless, if auditing is not established, then the whole environmental management system is weakened.

Three levels of auditing are used in Courtaulds to evaluate and improve health and safety management:

(1) behavioural audits — these are informal, activity-based, and involve as wide a range of employees as possible;

(2) compliance audits — these are formal, and performed by experts (the business safety and environmental advisors) who fully understand the process and the relevant legislation;

(3) system audits — these are performed by CHESS, and evaluate how the overall system is functioning.

This approach is now being applied to environmental auditing within Group businesses. While the last two levels of audit are most generally recognized as the main approaches to management system auditing, the first level can have considerable benefits to an organization, and where possible should also be adopted. It will probably not be practical to include all employees in the auditing of a manufacturing facility, but the system could incorporate line and shift managers and supervisors, for instance. This will help to raise environmental awareness across the business at all levels, and help to dispel any impressions that auditing is a punitive approach to identifying problems.

CONCLUSIONS

Although the principles underlying a systematic approach to environmental management are relatively straightforward, the practicalities of implementing such a system present a number of pitfalls to industry. On the basis of admittedly limited industry-wide experience to date, the key issues appear to be:

• making the commitment, which requires an understanding of needs and benefits, scope and resources;

• identifying the missing components in the cycle of management, and then closing this loop. Once this has been done, and an objective of continual improvement is established, then the system should be self-perpetuating and improving;

- auditing effectively;
- involving everyone — keeping them informed and committed;
- integrating health, safety, environmental and quality management. This is a laudable long-term goal, but to get environmental management started it is best tackled as a separate initiative.

It has been said that BS7750 can be made to fit industry, but that it will be difficult to make industry fit BS7750. This is very true, and as a consequence most companies will choose to mould the general specification for an environmental management system to suit their needs. Within the UK alone, industrial management incorporates a wide range of different cultures, styles and systems. The specifics for implementation in each case will necessarily need to reflect this diversity.

8. DEVELOPING A SYSTEM TO MEET BS7750 AT A MANUFACTURING SITE

Pauline Fawcett and John Nobbs

For many years British Nuclear Fuels plc (BNFL) has placed a high degree of emphasis on the control and management of its discharges to the environment, particularly those of a radioactive nature. As part of this emphasis, predictive models for assessing impact were developed and have become familiar, as have effluent reduction strategies where these have been necessary.

In the past, for a variety of reasons, more emphasis was placed on management of radioactive discharges. BNFL, like other major companies, has recognized the need to address broader environmental issues in a systematic way. This has led to a widening of the focus from radioactive discharges alone. Adoption of a systematic approach is seen as compatible with other contemporary quality moves — for example, the implementation of Total Quality Management (TQM) philosophy and existing Quality Assurance (QA) requirements (similar to ISO9000) placed on the industry.

A company environmental policy statement was developed and issued at the beginning of the 1990s. Initially 'environmental champions' or 'co-ordinators' developed and implemented the statement at the divisional level (BNFL has four operating locations and a number of divisions) and formulated local action plans. There were some difficulties with this approach, however, as co-ordinators had other duties and the responsibility for implementing action plans and the authorizing of expenditure was not always clear.

The introduction of BS7750 — providing a framework against which existing management systems could be tested, with the possibility of external certification of the management system to a national standard — led to considerable interest within the company. The Company's Health, Safety and Environmental Committee (CHSEC, a top level management committee) decided to carry out a pilot study to assess the existing management system for environmental matters against the requirements of BS7750, to identify which areas would need to be strengthened.

An additional reason for carrying out the pilot study was to support the British Standards Institution (BSI) 'pilot assessment programme', since BNFL was a participant — as an observer organization — in the Electricity Sector

Working Group of this programme. BNFL Fuel Division, a 220-acre site employing some 3000+ people at the time, was selected for the pilot study.

BNFL FUEL DIVISION PILOT STUDY
A study team was assembled, comprising the following members drawn from the Fuel Division and the Corporate Centre:
- Fuel Division QA Manager (Team Leader);
- Head of Corporate Health, Safety and Environmental Audit Group;
- Senior QA Engineer, Corporate QA;
- Environmental Advisor, Corporate Health and Safety;
- Senior QA Engineer, Fuel Division;
- Head of Corporate QA;
- Manager, Operations, Health Physics and Safety Department, Fuel Division;
- Environmental Officer, Fuel Division;
- Environmental Co-ordinator, Fuel Division.

An aerial view of BNFL Fuel Division. (Courtesy of BNFL.)

The study team progressed by regular team meetings, networking among team members and discussion interviews with, for example, line managers and managers responsible for certain environmental issues (waste management, land management and so on). The study was split into four stages:

STAGE 1: DEVELOPMENT OF CHECK LISTS
Check lists against BS7750 clauses and appendices were split into general and specific questions. The general questions covered divisional issues such as the policy and appointment of management representatives, whilst the specific questions covered procedures and particular records for types of activities — for example, discharge control and emergency management.

STAGE 2: IDENTIFICATION OF SIGNIFICANT ENVIRONMENTAL EFFECTS
Significant environmental effects were identified in the completed 'check lists', which were completed using discussion, previous experience of the team members and documentation. Visits to plant areas were not conducted. For the purposes of the pilot study, the team developed a pragmatic definition of 'significant', this being:
- obvious environmental effect;
- generating public concern;
- adversely affecting local community;
- adversely affecting employees;
- high usage of resources.

The team effectively made a judgement on these five matters for any effect identified.

STAGE 3: ASSESSMENT OF THE EXISTING MANAGEMENT SYSTEM

STAGE 4: COMPILATION OF CONCLUSIONS AND EXECUTIVE SUMMARY

As BS7750 was quite new at the time of the study, it was inevitable that the team identified concerns about interpretations or definitions, particularly focused on phrases such as 'continuous improvement' and that overused workhorse 'significant'. Feedback to BSI was through the company's involvement in the Electricity Sector Working Group.

The main conclusions from the pilot study were that:
- adoption of BS7750 would complement and systematize the environmental work of the company;

• existing divisional arrangements did not comply fully with BS7750, primarily in the areas of environmental training, effects registers and auditing;
• no new plant would be needed, but additions and alterations to existing management procedures would be required;
• costs would be critically dependent on the way in which 'continuous improvement' and 'significant effects' were interpreted by auditors, and the extent of demonstration of suppliers' environmental credentials and commitment.

The findings were reported to the CHSEC at Corporate Centre and a decision subsequently taken that the whole company would proceed to be in a position to apply for BS7750 certification by March 1996.

MEETING BS7750 REQUIREMENTS: FUEL DIVISION'S EXPERIENCES

Fuel Division decided to aim to be in a position to apply for certification by April 1995, which was considered to be a tight, but nevertheless achievable, target. In order to achieve it, the division set up a Key Improvement Team whose objective was to establish a comprehensive management plan to allow certification to BS7750. Specific tasks included establishing how a Significant Environmental Effects Register should be set up and maintained, how objectives and targets could be set, and the methodology/procedure by which an environmental management system could be implemented and maintained.

The identification of significant environmental effects proved to be one of the most onerous tasks. This was due to a number of reasons, considered in more detail later in this section, including the size and diversity of the operations, the amount of information to be handled, the assessment of such information and indeed the development of the evaluation technique itself to identify 'significant effects'.

Taking each of these points in turn, Fuel Division is a large industrial site which operates chemical plants for the purification of uranium from uranium ore concentrates and its subsequent manufacture into nuclear fuel. In addition to the chemical plants there are also machining and canning operations, and a number of services — engineering and instrument workshops, medical and occupational health facilities, laundry and restaurant facilities to name but a few.

With such a large and diverse site the first stage was to decide what information needed to be gathered and how it should be obtained. Initially, a list was compiled of the type of environmental effect to be addressed and the scope of the information required. This list specified all the areas to be covered and included such things as environmental policy requirements, discharges from process plant, the effects of transport, the effects of purchase of raw materials and so on.

The next stage was to decide how to obtain all the necessary information. From discussions it was agreed that most of the information would have to be collected via questionnaires issued to plant managers and managers of specific activities — for example, transport and purchasing. Whilst it was recognized that this may not be the ideal way, it was considered to be the most rapid and cost-effective solution. All returns were independently reviewed by experienced staff with personal knowledge of the process or activity, and against current databases (for example, COSHH) to ensure all items and activities had been covered. They were also reviewed against 'total site discharges' as a means of double-checking the information supplied on a plant-by-plant basis. In practice, approximately 50% of the returns were considered to be essentially complete with little or no additional information required. For the remainder, interviews were conducted with the respondents to obtain complete coverage. In addition to the questionnaire approach, a small amount of information was obtained from specialized personnel with a particular knowledge of an area in question — for example, land management practices.

All the data obtained for the register were placed on a computerized database, DATAEASE, on a number of forms created specifically for the exercise. These forms covered items such as:

- the type of environmental effect;
- where the effect occurs/arises;
- the size of the effect;
- what causes the effect;
- whether it is significant and the categorization of significance;
- reference to where the original data is held — whether it is on a questionnaire or elsewhere.

The use of a computerized database was considered beneficial for a number of reasons. These included the fact that the register is a 'live document' which is regularly updated and reviewed; it is more powerful than a paper system in that various fields on the form can be 'interrogated' and the data compiled in a variety of ways; it is available to a number of users simultaneously for a variety of different reasons. It may also be possible to track objectives and targets using this system, although this has not yet been considered in any detail.

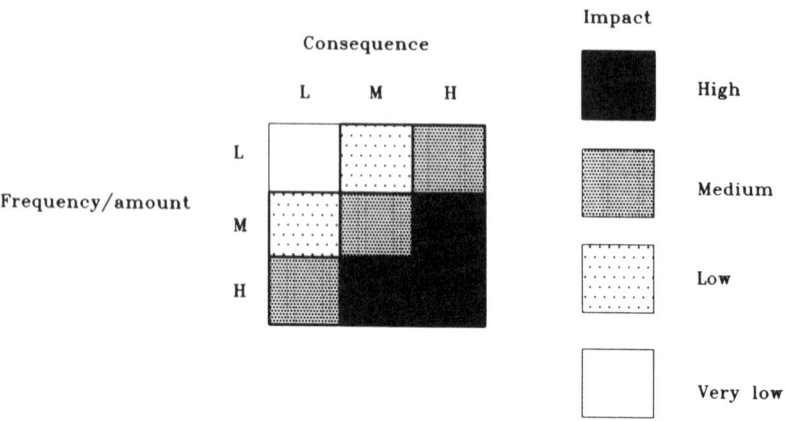

Figure 8.1   The Consequence/Frequency matrix.

Having reached the stage where all the information on 'environmental effects' was available on a database, the next stage was to carry out an evaluation of the environmental significance of the effects of the company's activities. Prior to starting this work, staff attended a number of presentations and seminars and had meetings with representatives from other companies undertaking the implementation of BS7750. On the basis of these experiences, coupled with the fact that the BNFL site is large, with a wide range of activities, it was decided that as a first pass a simple matrix would be used to assess significance.

For each entry on the effects register the judged impact (in terms of Low, Medium and High categories) was located on a Consequence/Frequency · matrix, the pro-forma of which is shown in Figure 8.1.

Whilst the approach was considered to be sound, several issues arose from this approach. These included such things as considering accidental or abnormal events separately from routine operations, using summed discharges on a site-wide basis where individual entries were recurrent from a number of different areas on site (for example, paper, asbestos, NOx emissions) and, perhaps most importantly, its limitation with respect to effects other than discharges.

Additionally, it was considered to be essential that the definitions of the terms used in the matrix were agreed, otherwise very different outcomes could be possible depending on the individuals or groups carrying out the assessment. It is also important in terms of the requirements of BS7750 and the fact that the register is a 'live document' and must be reviewed regularly. The following definitions were therefore agreed:

FREQUENCY/AMOUNT
- Low frequency/a small amount: Less than 100 te/a or less than once per week.
- Intermediate frequency/amount: Between 100 te/a and 1000 te/a or more than once per week but not daily.
- High frequency/a large amount: More than 1000 te/a or daily.

CONSEQUENCE
Consequence was a little more tricky to define clearly and the first approach was very much biased towards environmental and health standards to define consequence. It is recognized that no such standards exist for most environmental impacts (for example, paper usage, plastics disposal, visual effects and so on) and it will obviously be necessary to consider these issues somewhat differently. However, at the time of writing this issue had not been resolved.
- Low: Does not or is not expected to contribute more than 10% of a relevant 'environmental or health standard'. This could be an EQS, GLC, an OES, etc.
- Intermediate: Between the Low and High case.
- High: Is known or suspected to contribute more than 50% of a recognized 'environmental or health standard' or the relevant standard is close to being breached as a result of a number of activities with the company's activity being a contributor.

It was also recognized that with the information available at the time of the assessment it would be difficult to carry out a full quantitative impact assessment. It would therefore be necessary to make 'an expert value judgement' which would be made based on the many years' experience of the individuals making those judgements.

With regard to implementation of BS7750, it was decided at a relatively early stage that the environmental management system would be a part of the Total Quality System for the division. The present structure comprises a three-tier system:
- a divisional QA manual which defines the policy and establishes the system for ensuring compliance with quality standards;
- Fuel Division Management Procedures (FDMP) which define departmental responsibilities with respect to the business requirements which include, for example, such activities as planning, sales and marketing, management of environmental issues;

• local management systems which describe, if necessary, responsibilities detailed in the FDMP but concentrate on detailing any interfaces which may have an environmental impact.

Following the policy decision to comply with BS7750, the quality manual was amended slightly to reflect this requirement. The specific procedure dealing with the management of environmental issues was also modified to cover the responsibilities required by BS7750; these new responsibilities will automatically flow through to the departmental procedures.

With regard to organizational structure, several staff structure charts were considered and the merits of existing staffing arrangements versus the setting up of an 'Environmental Department' discussed. Two issues emerged as paramount. Firstly, the need for the Environmental Manager as defined by BS7750 to be at Senior Manager level to have the necessary authority to ensure implementation. Secondly, the need to define responsibilities clearly. It was decided that existing staffing arrangements would remain in place, which resulted in the detailed responsibilities and authorities for environmental management being described in the Fuel Division Management Procedures. In practice, it also requires good communication between all personnel with a specific environmental responsibility. Finally, one suggestion that emerged was the use of 'Environmental Representatives' along the lines of existing 'Safety Representatives', as an aid to management in maintaining day-to-day vigilance on environmental matters.

SUMMARY OF ACHIEVEMENTS TO DATE

The pilot study carried out in 1993 concluded that adoption of BS7750 would complement and systematize the environmental work of the company, but that some additional arrangements were necessary. For Fuel Division the major additional requirements included the need for more environmental training, the use of environmental audits and the compilation of an effects register. Since that time a team has been set up to establish a comprehensive management plan which will allow Fuel Division to apply for certification to BS7750. To date, the policy has been revised, a full register of all potentially significant effects has been compiled on a database, a method of assessing significance is being developed and a management system has been established to reflect the requirements of BS7750. It is intended that Fuel Division will apply for certification in August 1995 following completion of the programme and an internal audit.

9. A PRACTICAL APPROACH TO IMPLEMENTING BS7750 IN THE CHEMICAL INDUSTRY

Ken Jordan

Site Gillingham is part of the business unit Polymer Chemicals, within the chemicals group of Akzo Nobel. The site was officially opened as Novadel Ltd in January 1938 for the production of white lead, associated paint products and additives for the flour milling industry.

The 18-acre site on the banks of the River Medway in Kent, UK, is one of four major organic peroxide producing locations within the EU. It is operated by Akzo Nobel's Chemical Group, and employs some 140 personnel.

Five major manufacturing units, plus several minor units, produce speciality chemicals including organic peroxides for the plastics and rubber industries and a monomer for the production of an organic glass for the optical industry. The site exports 95% of its manufactured tonnage from the UK, with Europe taking two thirds of the total production. The Middle East and Far East are fast-growing markets; sales to these areas have increased from 10% of the total sales to 20% in the last three years.

The site also serves as a UK distribution centre for other Akzo Nobel products manufactured outside the UK. The site was registered to ISO9002 in June 1990. A commitment to achieve the 'Investor in People' award was signed in June 1993, with registration expected in mid-1995.

The site was assessed for the new environmental management standard BS7750 in March 1992, with registration granted in March 1994.

INTRODUCTION

When planning the introduction of BS7750, Akzo Nobel used the implementation circle from the standard. The company started off with a commitment when the standard was launched in March 1992 and in March 1994 received accredited certification to BS7750. Akzo Nobel has a corporate policy to introduce environmental management systems in all of its plants; when Akzo Nobel systems were compared with BS7750, the standard was, at that time, superior. Nevertheless, an integral part of the Akzo Nobel environmental management system was a Dutch Government environmental questionnaire. This questionnaire, modified to suit the particular circumstances of Akzo Nobel operations, was used to carry out the initial environmental review.

INITIAL REVIEW

The site was split into 25 areas and different people were assigned to complete the modified questionnaire for these areas. The results from the questionnaires were then correlated and summarized to give a base line for the site's environmental performance.

Several issues and objectives were identified from this initial review:

• it was clear that all environmental legislation was being met; however, it was difficult to quantify environmental performance;

• it was decided to compile a site environmental manual which could be updated annually so that detailed records of performance for air, land and water discharges could be maintained;

• it was also obvious, at an early stage, that everybody should receive basic environmental awareness training;

• the following policy statement was issued as a result of the initial review.

Akzo Nobel Chemicals Limited, Site Gillingham will establish and maintain an environmental management system to fulfil the requirements of BS7750, 1994. The system will ensure compliance with current UK environmental legislation and best available practices, and achieve a balance between economic, social and environmental responsibilities. We are committed to avoiding damage to the environment by any of our actions or operations.

Site Gillingham is dedicated to continual improvement of environmental performance, and efficient use of resources, which will be achieved by setting and ensuring successful implementation of environmental objectives. This policy will be publicly available and will be understood, implemented and maintained by all levels in the organization.

The company environmental policy allows for the management to manage the business to achieve a balance between economic, social and environmental responsibilities. It is clear that management plans to improve environmental performance, as required by BS7750, and individual elements of the plan, have different time scales and need to be reviewed regularly to ensure continual improvement.

Akzo Nobel Chemicals Ltd offices at the site in Gillingham.

After publishing the policy, 140 people on the site received environmental awareness training. This half-day training session included general environmental awareness information, an explanation of BS7750 and a group exercise to show how people could personally contribute to improving the site's environmental performance. The environmental awareness training took place in about 10 sessions on site. Feedback from these sessions was brought together and used to generate projects which were incorporated into the site's environmental targets and objectives. It was important at this stage that everybody on site was aware of the environment and feels part of managing the environment. To reinforce the site's commitment to the environment, a site environmental magazine is published every six months to communicate performance and progress of projects. It was interesting to note that the interest in environmental issues was much greater than in quality training.

MANAGEMENT RESPONSIBILITY
As with most major chemical companies, people have already been made responsible for environmental issues and this responsibility is now regarded with

equal importance to health and safety on the site. There are, however, prime environmental responsibilities where the Health, Safety & Environmental (HS&E) Manager is responsible, including legal waste disposal procedures and the maintenance of current legislation. The Production Manager is responsible for operating the plants and environmental protection equipment to meet environmental standards. The Product and Quality Control (PQC) Manager is responsible for monitoring environmental performance. These are clearly defined responsibilities.

LEGISLATIVE REGISTER

To keep up to date in the area of legal regulations, the Barbour Index is used. This is a professionally prepared system which gives a quarterly review of updates or changes in legislation (for further information contact Barbour Index plc, New Lodge, Drift Road, Windsor, Berkshire SL4 4RQ). The HS&E Manager is responsible for ensuring that any changes to current legislation affecting the environment are communicated to the responsible managers on site.

REGISTER OF SIGNIFICANT EFFECTS

The procedure for the generation of the register of effects was, in principle, very simple. The first step was to generate an exhaustive list of all effects, then to assess their significance and finally to produce a register of significant effects.

When generating the exhaustive list of effects, consideration was given to all raw materials, final products, processes, plants and buildings for their effects on air, water and land resource usage and nuisance, taking into consideration both normal and abnormal conditions. This list was then screened against certain criteria which define a significant effect. Akzo Nobel chose four different criteria for significance, ranging from an insignificant effect such as spilled water, to a significant effect when under normal operations the plant or process is operating illegally. The exhaustive list of effects could then be given an environmental significance.

Once this list of significant effects had been prepared, targets and objectives based on the environmental policy could be set for improvement.

MANAGEMENT PROGRAMME

Having identified the significant effects, Akzo Nobel uses a project management system to ensure that these environmental targets and objectives are met. Significant effects can be combined to form individual projects which can be assigned different targets, objectives and time scales for implementation. The site management team reviews projects quarterly to ensure that targets and objectives are being met and that the appropriate priority is assigned.

The site's environmental objectives incorporate the significant effects from the environmental register and ideas identified by the feedback from the site environmental awareness training.

SYSTEM INTEGRATION AND AUDITING

The site's quality management system and environmental management system structures are very similar and the requirements of BS7750 have been integrated into the site management system. This has many advantages; internal audits are not carried out differently for ISO9000 and BS7750, and audits are not carried out twice. Internal auditor training is seen as an important part of maintaining the system; however, the internal audits are basically the ISO9000 system audit. A management system internal auditor requires training — both ISO9000 auditor training and practical environmental training sessions for specific site needs. Although system auditing is all that is required by BS7750, the next logical step is to carry out performance auditing, which will be required for Eco-Management and Audit scheme (EMA) certification. In 1994 Akzo Nobel (Gillingham site) produced its first site environmental report based on the Conseil Européen de l'Industrie Chimique (CEFIC) guidelines. This quantifies the company's performance and makes the site policy and objectives publicly available. It is intended to publish a report every year, which is intended to meet the reporting requirement under EMA.

CONCLUSIONS

Since the launch of BS7750 in early 1992, Akzo Nobel has implemented the system step by step and finally received accredited certification in March 1994. The time taken to implement the system on the site has been estimated as three man years. It is important to note that the objective of a 10% saving in energy use, if achieved, will pay for the complete project.

10. ENVIRONMENTAL MANAGEMENT — AN EXAMPLE FROM THE FINE CHEMICALS SECTOR

Rodney Perriman

Zeneca Limited's approach to the environmental aspects of its activities is that protection of the environment must be an integral consideration, and a line management responsibility, for all of its operations. In common with most of the UK chemical industry, Zeneca is adopting a common system for the management of safety, health and environment (SHE). Since its demerger from the former ICI Group in 1993, Zeneca has given very high priority to establishing its SHE management system throughout the Group. This chapter describes the key elements of the system and its implementation.

Compliance with the requirements of integrated pollution control (IPC) under Part I of the Environmental Protection Act 1990 (EPA90) is one of the environmental considerations that the SHE management system must address in the UK. The need to demonstrate best practicable environmental option (BPEO) poses particular challenges for the development of new processes in the manufacture of fine chemicals. The second part of this chapter outlines an approach to these challenges.

THE COMPANY

Zeneca is a leading international bio-science group, developing and creating innovative products that meet customers' needs worldwide in the areas of:

- health care (Zeneca Pharmaceuticals);
- nutrition (Zeneca Agrochemicals and Zeneca Seeds);
- speciality products (Zeneca Specialties).

Zeneca has global coverage; it manufactures in 25 countries and its products sell in over 130 countries. 30,000 employees worldwide contribute to an annual turnover (1993) of £4.44B.

THE SHE MANAGEMENT SYSTEM

The group's overall system for managing SHE across a wide and diverse range of activities is represented in Figure 10.1 on page 110.

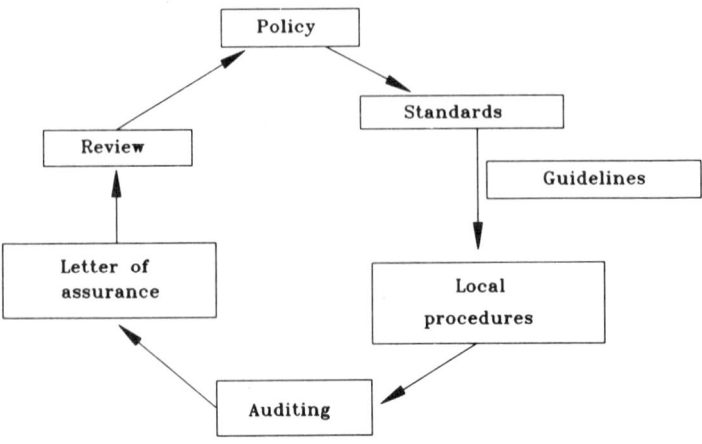

Figure 10.1   The Zeneca SHE management system.

POLICY

In the area of SHE the company's policy is a public statement of its commitment. The board of Zeneca has set down its basic requirements in the form of a policy for SHE. This policy clearly states Zeneca's position on SHE matters and all its employees are required to comply with them:

'In pursuit of its business objectives, it is Zeneca's policy to manage its activities to give benefit to society ensuring that:
• they meet all relevant laws, regulations and international agreements;
• they are conducted safely, protecting the health of all employees and all persons who may be affected;
• they are acceptable to the community at large;
• their environmental impact is reduced to a practicable minimum at an acceptable cost to the Group and society.'

STANDARDS

The broad statements of commitment in the SHE policy have to be interpreted in relation to all the activities carried out by the company. For this purpose, and to ensure consistency in meeting the aims of the policy throughout the Group, the Zeneca board, in consultation with chief executive officers of the businesses, has set down 19 Group SHE standards which must be complied with by all managers throughout Zeneca. These standards are concise statements of the basic management requirements and are applicable to all of the activities in which Zeneca participates. The areas covered by the 19 standards are:

- SHE commitment;
- management and resources;
- communication and consultation;
- training;
- material hazards;
- acquisitions and divestments;
- new plant and process design;
- modifications and changes;
- SHE assurance;
- systems of work;
- emergency plans;
- contractors and suppliers;
- environmental impact assessment;
- resource conservation;
- waste management;
- soil and groundwater protection;
- product stewardship;
- SHE performance and reporting;
- auditing.

GUIDELINES

Line managers are responsible for ensuring that all the activities under their control are carried out in accordance with the general requirements of the Group standards. They must also take account of any particular requirements of their business and all local legal and regulatory requirements. Such details cannot be determined centrally. However, the Group has a wide range of experience and skills amongst its managers and SHE professionals throughout its businesses, territories and corporate functions. This collective knowledge is made available to line managers in the Group good practice guidelines. These provide guidance to managers on the key principles which should be incorporated into local procedures and provide advice on how the principles should be applied. They also give references to other relevant documented sources of information.

LOCAL AUDITABLE PROCEDURES

All locations must have their own local procedures which set out the local arrangements and requirements for carrying out work in a safe, healthy and environmentally sound way. Such procedures are mandatory at the location. The extent of documentation is a matter for local judgement in relation to the hazards and risks associated with the activity being managed.

Auditing

Auditing is a systematic examination of activities and procedures which provides assurance to managers that the systems in place are adequate, and that activities are being carried out in accordance with the requirements of Group policy and standards and all local regulatory and business requirements. Auditing alerts them to any improvements needed. Zeneca has adopted a Group-wide approach for training selected managers as SHE auditors and all businesses are implementing audit programmes worldwide.

Letter of Assurance

Information from audits is included in the annual Letter of Assurance, which is delivered to the executive board by the chief executive officer of each business following a review of SHE performance. The document outlines the extent of compliance with company standards and indicates areas, plans and time-scales for improvements.

Review

In the light of the Letters of Assurance and reports on SHE performance, the executive board can assess the extent to which the policy is being fulfilled and consider the need for changes to further the continuous improvement of performance.

PROCESS DEVELOPMENT

A key feature of fine chemicals production is the frequent introduction of new products. The need to demonstrate BPEO poses particular challenges for the manufacture of fine chemicals. Section 7 of EPA90 requires that the conditions in an authorization for a prescribed process shall satisfy the objective that the process represents 'the best practicable option available as respects the substances which may be released' to the environment. In practice this means that the person submitting an application to Her Majesty's Inspectorate of Pollution (HMIP) has to demonstrate that the process described in the application is the BPEO for that process in that location.

In April 1994 HMIP published a consultation paper[1] proposing a procedure for 'Environmental, Economic and BPEO Assessment Principles for Integrated Pollution Control'. The chemical industry had considerable reservations about several aspects of HMIP's proposals, particularly the concept that a variety of environmental effects, resulting from releases to air, water and land from a process, might be expressed as a single number — and that number used as a measure of the overall environmental impact of the process.

An alternative approach has been suggested, which would form an integral part of the design of new processes and substantial modifications to existing processes. It would involve:

- identifying the releases from the proposed process;
- determining the resulting effects on the environment;
- carrying out the same procedure for other practicable technical options for the process;
- comparing the overall environmental impacts of the proposed process and the practicable options;
- comparing the costs of the proposed process with the costs of the practicable options;
- assessing whether the proposed process had the least adverse environmental impact of all the options, or whether the additional costs associated with any options of lower environmental impact than the proposed process were reasonable in relation to the reduction in environmental impact.

The comparisons and assessments involved in this procedure could not be done by allocating and summing numerical values based on disparate environmental effects to produce an 'environmental index' for a process. It would require informed judgement to decide on the relative differences between the options and to determine whether any additional costs of an option with lesser environmental impact were reasonable in relation to the environmental improvement obtained.

All the information used to carry out this procedure would be fully documented, as would the reasons for the selection of the proposed process in relation to any of the practicable alternatives.

The benefit of this procedure is that the reasons for making the final process selection on environmental grounds are apparent and can be clearly documented as an audit trail — as recommended by the Royal Commission on Environmental Pollution in its *12th Report*[2]. The value judgements, which cannot be avoided in any procedure which tries to balance disparate environmental (and other) factors, can be clearly explained — they are not obscured, as they would be by the use of pre-determined pseudo-scientific factors for environmental impact, such as indexes compiled from ratios of emissions or environmental concentrations to arbitrary environmental quality limits.

This method should be equally applicable for manufacturing processes where there are few practicable technical options, as is often the case for large scale commodities production — for example, there are only two established routes for the manufacture of titanium dioxide — and for the more complex process route selection procedures in the synthetic organic chemical industry.

The following paragraphs describe the method in more detail as the procedures would be applied for synthetic chemical manufacture. The method is shown in diagrammatic form in Figure 10.2.

The principles and generic procedures involved are likely to be appropriate for any process subject to IPC. In many cases, their application would be simpler because other industries do not have the multiplicity of potential process options that have to be considered in selecting the manufacturing route for a new synthetic chemical.

BPEO ASSESSMENT PROCEDURE FOR CHEMICAL MANUFACTURE
Before starting the process design the following data are generally well defined from commercial considerations:
* quantity of substance to be manufactured;
* quality of substance to be manufactured;
* range of acceptable manufacturing costs;
* time-scale to beneficial manufacture.

STAGE (1): GENERATE SYNTHESIS ROUTE OPTIONS
For most organic molecules there are many potential chemical pathways by which they could be produced and the number increases with the complexity of the molecule. These pathways are generally based on theoretical considerations and many will not have been evaluated by experiments. Each pathway produces a particular combination of wastes and potential releases to the environment,

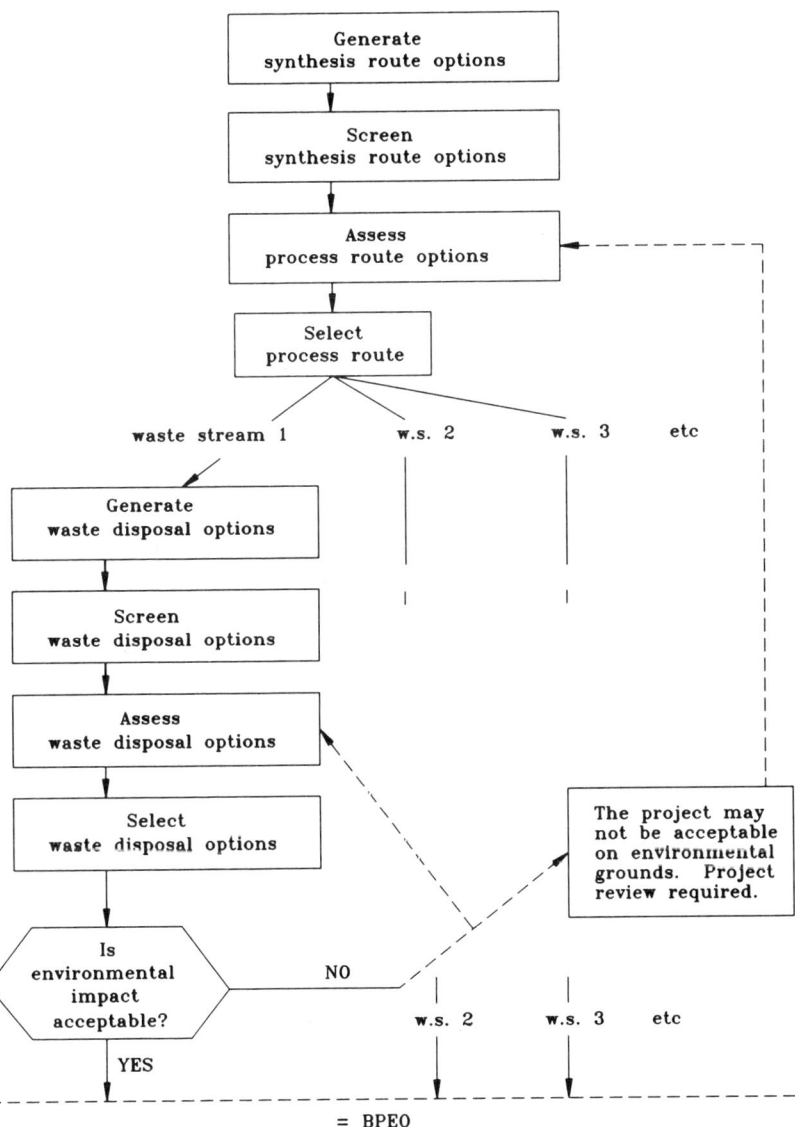

Figure 10.2   BPEO assessment procedure for manufacture of a speciality organic chemical. (Note: each waste disposal option must be BATNEEC.)

both in quantity and type. For a complex molecule a comprehensive review at this stage may typically identify 40–100 pathways.

STAGE (2): SCREEN SYNTHESIS ROUTE OPTIONS
It is not practicable to develop such a large number of synthesis route options to the stage of preliminary process design, nor is it necessary. The normal practice is to reduce the number of options for further consideration to between three and about six by the application of technical assessment, professional judgement and very limited experimentation. Factors to be considered include:
• technical viability;
• SHE issues;
• cost;
• likely process development time.
    The short-listed synthesis route options will be those which are judged to be worth examining in a more detailed study as process route options.

STAGE (3): ASSESS PROCESS ROUTE OPTIONS
The procedure described here is concerned with the environmental assessment of process routes for each of the short-listed synthesis routes identified in stage (2). Concurrent with the environmental assessment there will also be assessments of other aspects such as:
• capital and operating costs;
• safety;
• health;
• technical development needs;
• operability.
    The environmental assessment of process route options at this stage will normally be based on:
• a preliminary process design/flowsheet;
• reaction stoichiometry;
• estimated reaction yields;
• available toxicological and ecotoxicological information
    Using this information, each process route option is assessed in terms of parameters which are indicative of the environmental impact of the process. By way of example, the following list of parameters may be used at this stage in the design of a process for manufacturing a new speciality organic chemical:
• Theoretical amount of waste produced per unit of product produced if reaction

yields were 100%. (This indicates the amount of waste which will not be reducible by process technology improvement.)

• Forecast amount of waste produced per unit of product based on expected reaction yields. (This indicates the amount of waste which is likely to arise but which may be reduced by process efficiency improvements.)

• Number of recycle streams assumed in the proposed process. (Experience shows that recycle streams envisaged at this stage are often found to be impracticable as the detailed design is developed. They may then become additional waste streams.)

• Number of difficult substances handled — for example, listed compounds or ones with severe ecotoxicological properties.

• Amount of special or hazardous waste produced.

• Total organic carbon in effluent streams.

• Total dissolved solids in effluent streams.

• Nitrogen in effluent streams.

• Phosphorous in effluent streams.

• Number of volatile organic compounds handled.

• Energy consumption.

• Number of identifiable circumstances where an unplanned release would have severe effects.

The list of parameters is chosen to be relevant to the type of process under consideration and the environmental concerns of the site. Other parameters might be relevant and important for other types of processes in other locations — for example, acid gas emissions, greenhouse gas emissions, persistent and toxic organics in effluent, heavy metals in effluent and so on.

For each process route option the appropriate value of each parameter is estimated.

STAGE (4): SELECT PREFERRED PROCESS ROUTE
The numerical values against each parameter are ranked across the process route options and an overall ranking of the options is derived. This procedure is illustrated in Table 10.1 (see pages 118 and 119) for a hypothetical but realistic example of a new speciality chemical.

The route with the highest overall ranking is likely to be the best environmental process route option, but that conclusion must not be accepted automatically. Simple addition of rankings against each parameter to produce the

117

TABLE 10.1
Comparison of process routes

	Route						
Indicator	1	2	3	4	5	6	7
Theory ELF[1]:							
ELF	2.72	4.6	2.92	4.68	4.58	3.38	2.4
Ranking[2]	2	6	3	7	5	4	1
Actual ELF:							
ELF	4.3	6.94	4.52	6.12	7.7	5.02	7.94
Ranking	1	5	2	4	6	3	7
Dependence on recycles:							
No. of recycles	7	7	7	6	11	5	7
Ranking	3	3	3	2	7	1	3
Nature of substances handled[3]:							
No.	2	4	3	4	5	6	6
Ranking	1	3	2	3	5	6	6
Hazardous waste produced[4]:							
tpa	2800	2900	2900	1000	150	500	1200
Ranking	5	6	6	3	1	2	4
TOC[5]:							
tpa	210	300	300	220	460	180	840
Ranking	2	4	4	3	6	1	7
TDS[6]:							
tpa[3]	450	5700	450	8700	13,500	7050	7200
Ranking	1	3	1	6	7	4	5
N&P[7]:							
tpa	180	360	360	220	560	560	1380
Ranking	1	3	3	2	5	5	7

TABLE 10.1 (continued)
Comparison of process routes

	Route						
Indicator	1	2	3	4	5	6	7
VOCs[8]:							
No.	4	3	3	7	6	5	5
Ranking	3	1	1	7	6	4	4
Unplanned releases risk[9]:							
No.	2	4	4	4	6	6	7
Ranking	1	2	2	2	5	5	7
Overall:							
Score[10]	20	36	27	39	53	35	52
Ranking	1	4	2	5	7	3	6

Notes:
[1] ELF = environmental load factor, the quantity of waste per unit of product produced
[2] ranking is 1 (the best) through to 7 (the worst) for seven route options
[3] nature of substances handled is the number of substances involved which are of high environmental concern
[4] hazardous waste produced is the forecast amount of special waste
[5] TOC = total organic carbon in the effluent
[6] TDS = total dissolved solids, generally inorganics, in the effluent
[7] N&P = total nitrogen and phosphorous in the effluent
[8] VOCs = volatile organic compounds
[9] unplanned releases risk is the number of identifiable situations with potential high risk to the environment
[10] overall score is the total of the individual rankings

overall ranking implies that the environmental impacts of each parameter are equally important. This is not necessarily so.

Before making the final selection on environmental grounds, the ranking under each environmental impact parameter must be examined to ensure that one severe and important impact is not overlooked simply because it has been counterbalanced by several low impact parameters which may be of lesser importance in the particular site circumstances. At this point consideration should also be given to the precise nature of some of the parameters — for example, the classification of hazardous waste is very broad and a large tonnage of one type may have less potential environmental impact than a lesser tonnage of another.

In making the final selection of the best practicable route it is also necessary to take into account the results of concurrent assessments of process route options in terms of costs, technical feasibility and so on. Limitations of existing equipment may be a constraint, particularly in multi-stage batch processes which are typical of the fine chemicals sector. Other regulatory controls on manufacturing for some groups of licensed products may also limit freedom of choice. These practical considerations may lead to final selection of an option which is second or third best on purely environmental grounds. Again, the display of all the factors, as in Table 10.1, enables identification of the environmental consequences and how the balance between, say, environmental impact and costs and other factors has been determined.

STAGE (5): GENERATE WASTE TREATMENT AND DISPOSAL OPTIONS
Once the preferred process route has been identified, the next stage is to identify all waste streams from all stages of the process, as either minor releases (for example, storage tank vents) or major waste streams (for example, by-products, used catalysts and solvents, inorganic salts, distillation residues and so on).

For each of these major waste streams, all possible waste management options are identified, including options to reduce, reuse or recycle as well as end-of-pipe treatments and disposal.

STAGE (6): SCREEN WASTE TREATMENT AND DISPOSAL OPTIONS
In contrast to the sometimes large number of options which may warrant consideration at the synthesis and process route selection stages, the options for dealing with a particular waste stream are normally limited. Even so it is not practicable to carry out detailed design for every possibility. The list of possible options for each waste stream is screened, using a similar approach to stage (2), and reduced to two or three options for each waste stream. In screening the waste disposal options the waste management hierarchy of avoid — minimize — reuse — recycle — recover energy — dispose should be followed.

STAGE (7): ASSESS WASTE TREATMENT AND DISPOSAL OPTIONS

Each of the two or three waste management options identified in the previous stage is now assessed in more detail. For each option the releases to the environment are identified and quantified. Reference should be made to any relevant Chief Inspector's Guidance Notes (CIGNs)[3] to check on emission limits indicated as achievable using the best available technology not entailing excessive cost (BATNEEC) (see stage (10)). Where alternative options have releases of the same substances, a comparison of relative environmental impacts can be made simply on the basis of amounts released. Otherwise it will be necessary to tabulate the releases, possibly carry out dispersion calculations for releases to air or to water to estimate the levels of contamination in the local environment, and make comparisons with local environmental quality limits. (New guidance for carrying out this type of assessment is being developed by HMIP's Co-operative Development of Tools for Environmental Analysis project.)

Comparisons then have to be made between the environmental effects of each option to determine, for the particular site, which has the least overall environmental impact. This assessment is best done by a small team of people with knowledge of the environmental effects of the substances released and understanding of the priorities for the local environment. Considerations of wider environmental consequence may also be relevant when the releases include substances which contribute to long range effects and/or are subject to national emission limits — for example, sulphur dioxide.

The assessment must take account of recycling, reuse or waste treatment processes which take place outside of the boundaries of the proposed process for IPC authorization.

This procedure is repeated for each of the waste streams from the preferred process.

STAGE (8): SELECT PREFERRED WASTE MANAGEMENT OPTION

The option for each waste stream which has been identified in stage (7) represents the best environmental option for that waste stream.

The costs of each option examined at stage (7) are estimated to the degree of accuracy which is practicable at this stage. These would not be detailed engineering estimates, but preliminary estimates which indicate order of magnitude costs. The differences in costs between options are then compared with the differences in environmental impacts. Unless the least cost option is also the

121

option which is judged to have the least environmental impact, it will be necessary to decide whether any additional costs of an option with lesser environmental impact are reasonable:

- in relation to the economic viability of the project; and
- in relation to the environmental improvement obtained.

The latter assessment is, again, a value judgement and best done by a small team of people with knowledge of the environmental effects of substances released and understanding of the priorities for the local environment.

Other aspects of the waste management options — such as energy use, technical feasibility and so on — also influence the final selection of the best practicable waste management option for each waste stream. This includes checks on the compatibility and operability of the waste treatment methods as an integrated series of operations with the process itself. In some situations the existing site infrastructure and established waste treatment facilities serving other processes will also be important considerations in determining the most cost-effective combination of waste treatment methods.

STAGE (9): FINAL CHECK THAT ENVIRONMENTAL IMPACT
IS ACCEPTABLE

The preferred process route and associated waste management options identified by the above procedure are the BPEO for manufacturing the new product.

The environmental effects of the final releases to the environment from the process and its waste treatment systems must now be collated and documented as part of the preparation of the application for an authorization under the IPC requirements of Part 1 of EPA90. This stage should also be used as a final check that all the ultimate releases to the environment from the proposed process are acceptable in relation to all relevant environmental quality criteria for the site.

Since the procedure for selecting the preferred process and waste management options is designed to identify that combination (commensurate with economic and technical viability) which produces the lowest environmental burden, it is very unlikely that this final check assessment will indicate any unacceptable environmental impacts. But if any release is found to result in any breach of an environmental quality limit it is necessary to review the process route and/or waste management selection procedure, as indicated by the dotted line in Figure 10.2 (see page 115).

STAGE (10): BATNEEC

The detailed design and proposed operating procedures of the plant and equipment for the process route and waste management systems which have been selected as BPEO must also meet the current requirements of BATNEEC, as they apply to the process and site-specific circumstances of the project, in respect of minimizing and rendering harmless any releases to the environment. The series of CIGNs[3] published by HMIP provide guidance on these aspects.

STAGE (11): DOCUMENTATION

All the information used to carry out this procedure must be fully documented. These records are the source documents for compiling that part of the application for IPC authorization which demonstrates, to the satisfaction of HMIP, that the plant for which authorization is sought is the BPEO and that the best available techniques not entailing excessive costs will be used to minimize releases and avoid causing harm to the environment.

REFERENCES IN CHAPTER 10

1. HMIP, April 1994, *Environmental, Economic and BPEO Assessment Principles for Integrated Pollution Control; Consultation Document.*
2. Royal Commission on Environmental Pollution, February 1988, *12th Report: Best Practicable Environmental Option* (HMSO).
3. Regulating Systems Division of HMIP, *Chief Inspector's Guidance to Inspectors — Industry Sector Guidance Notes and Process Guidance Notes* (HMSO).

11. TENPRO — AN INTEGRATED SOLUTION TO THE DEMANDS FOR ENVIRONMENTAL ACCOUNTABILITY

Andrew Beedle

In process and manufacturing industries, environmental accountability is now seen as a public obligation and a business imperative. Regulatory standards are becoming increasingly stringent whilst the public debate about the environment has become better informed and shows growing concern. The move towards reduced environmental impact and cleaner manufacturing is now seen by the business community as a matter of prestige with a direct impact on market share and the bottom line.

THE MARKET FOR CONTINUOUS ENVIRONMENTAL MONITORING SYSTEMS

A series of key market drivers can be identified as a direct result of growing public concern. Public pressure exerted on government and consequently on the regulatory authorities has resulted in the passage of a raft of legislation at European and national levels. In the UK, the concept of integrated pollution control embodied in the Environmental Protection Act 1990 and its associated legislation has given rise to the strongest market driver — legislative compliance. Public opinion, a second key market driver and the driving force behind much of the recent legislation, also exerts considerable influence on the market, forcing manufacturing companies to take account of their environmental performance and their public image. This, in turn, has resulted in the appearance of a third key driver — corporate pressure. Many companies, having recognized the bottom line benefits associated with reduced environmental impact and improved public image, are setting internal environmental standards which are often more stringent than those demanded by the regulators.

These factors have combined to compel manufacturers to quantify their effect on the environment and look for means of reducing their environmental impact. This has given rise to a number of emerging market trends including:

• an increasing demand for real-time environmental data;

- an increasing need to store securely, retrieve and manipulate large volumes of auditable environmental information;
- an increasing need to analyse and distribute environmental information throughout the manufacturing enterprise;
- a need to report environmental performance accurately — both within the manufacturing enterprise and externally — in order to demonstrate regulatory compliance.

THE BENEFITS OF REGULATORY COMPLIANCE AND REDUCED ENVIRONMENTAL IMPACT

The benefits of responding to the market drivers and satisfying these emerging trends are substantial. In financial and business terms they include:

- reduced waste and improved cost control;
- increased institutional and shareholder confidence;
- enhanced market share and/or margins;
- reduced insurance premiums.

In legal terms they include:

- the avoidance of litigation, fines, legal costs, management costs in dealing with legal actions, clean-up costs and further civil liabilities;
- the increased confidence of the regulators.

In terms of public image the benefits include:

- improved local relations and image;
- enhanced attractiveness as an employer.

THE TENPRO PROJECT

Having recognized the impact of the key market drivers and the benefits of regulatory compliance in the early 1990s, both ICI and Electricidade de Portugal (EDP) realized that there were, at that time, no tools available to help quantify their environmental impact and continuously monitor environmental emissions and discharges. Consequently, ICI, EDP and the Sema Group combined to submit a proposal to the European Commission in late 1991 to develop such a tool. This resulted in TENPRO, Total Environmental Protection, an 8.5MECU (£6.5M), 42-month project involving some 60 man years of research and development. The project aims to address the environmental needs of the process industries and is part of the 9BECU (£6.75B) ESPRIT III 'Industry and the Environment' work programme.

The result of this 60 man years of effort is an environmental supervisory system for managing emissions and discharges from industrial plant. Although designed primarily for the process and power industries, TENPRO enables the more efficient management of emissions and discharges across a wide range of industry sectors, thereby reducing operating costs whilst increasing productivity.

TENPRO has three core modules:

• the Measurement Gateway provides the interface to the real world, allowing the continuous collection of environmental data from a wide variety of external sources including measurement sensors, plant control systems, laboratory systems and business systems;

• the Environmental Accounting module allows the continuous recording of key environmental measurements and the monitoring of compliance against regulatory and other limits;

• the Model Server allows the seamless integration of user-developed models into the TENPRO environment, providing open access to the environmental information maintained within the system and interfaces to distributed applications across an enterprise.

These core modules directly address the emerging market trends, with the design of the system strongly influenced by established user needs. A system overview is given in Figure 11.1 on page 128.

TENPRO is based on an open, distributed, object-oriented architecture and has been designed to be independent of underlying platform hardware and to be totally scaleable. This allows the implementation of individual applications at plant, site, regional and corporate levels, enabling the total integration of environmental information across a manufacturing enterprise. For example, if TENPRO were to be applied on individual plants in an enterprise, compliance reports for those plants could be automatically generated with site-related environmental information being passed to a site-wide system. This information would then be available for, as an example, site-wide performance improvement initiatives and reporting requirements. Environmental information from site systems could then be consolidated at a corporate level to provide, for example, total enterprise emissions and discharge information for the company's annual report.

The system makes use of best-in-class components with the ability to replace individual modules as the software and technology is developed. This ensures 'future proofing' in that, as soon as an individual component — whether

it be the information repository or the means of collecting environmental information — is superceded, it can be removed and replaced with the current best-
in-class offering.

The system provides a secure, auditable repository for collected environmental information with audit trails maintained for all transactions. Thus, if
a series of collected data points are unreliable due to, for example, an instrument

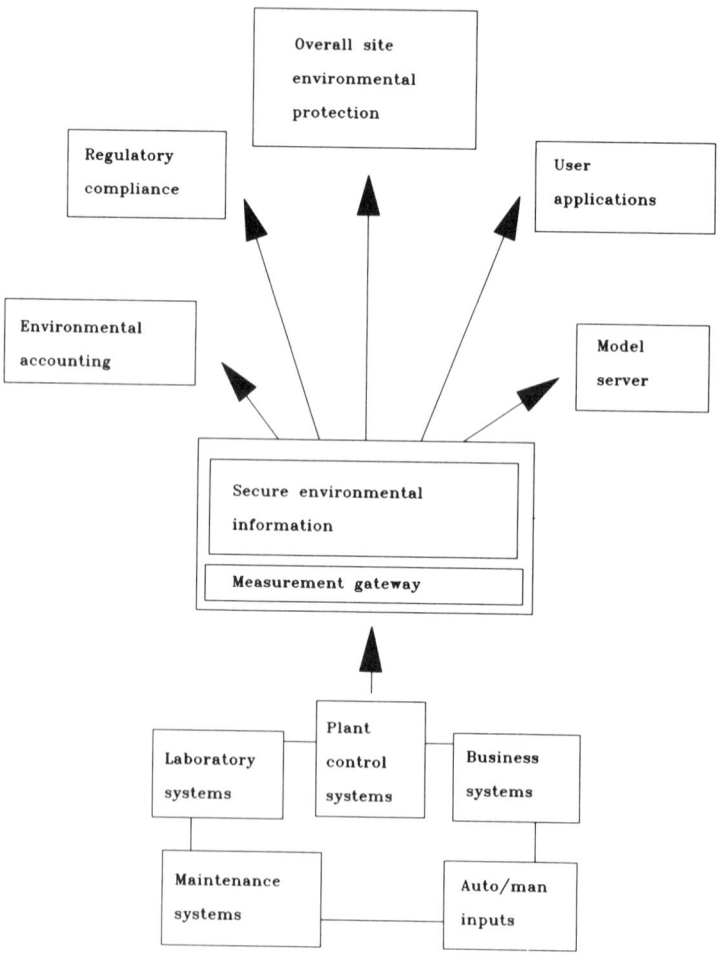

Figure 11.1   TENPRO system overview.

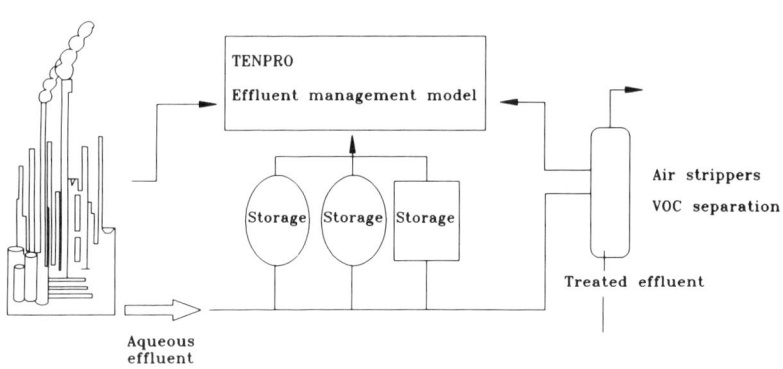

Figure 11.2   TENPRO effluent management system.

malfunction, the unreliable data can be corrected or replaced as appropriate but only if the individual responsible is named and a reason provided.

The combination of the functionality provided in the core modules and the software development environment provided by the TENPRO platform enables the development of user-specific applications and the support of other business processes such as concurrent engineering, product and process life cycle analysis and environmental impact assessment.

APPLYING TENPRO TO INDUSTRY
Industry applications of TENPRO are currently being developed in the chemicals and power sectors.

In the chemicals sector, TENPRO is being applied as part of a major environmental improvement project in North West England. The system will be used to monitor continuously emissions and discharges and the resultant position against site consents, in addition to improving manufacturing asset efficiency and performance through the development of specific, user-defined production models.

Part of the environmental improvement project is the commissioning of a thermal oxidizer, subject to HMIP consents, handling gaseous effluent from a large, integrated chemical production site. This is illustrated schematically in Figure 11.2. Aqueous effluent is stored on site prior to volatile organic compounds being separated out and incinerated. Limited storage capacity and the cost of shutting down the thermal oxidizer mean that a delicate balancing act is required to ensure upstream production and effluent feed to the oxidizer is maintained without exceeding the fixed effluent storage capacity. A stochastic model, hosted by TENPRO, is currently being developed using parameters such as upstream production rates, treatment plant availability and local rainfall to provide management information such as current and projected effluent stock levels. The use of various 'what if' scenarios will allow management to gauge the effect of changes in production, rainfall and so on, and will enable more efficient effluent treatment and running of the thermal oxidizer and help ensure compliance with HMIP consents.

The HMIP consent also covers the size of the plume from the oxidizer stack. The plume should not be visible to local residents — to remove the plume, heat is input to the stack. However, depending on the ambient weather conditions — temperature, humidity and so on — the quantity of heat required to ensure a non-visible plume varies. A further model hosted by TENPRO and illustrated schematically in Figure 11.3 is being developed. This uses the ambient weather conditions and the composition of the gaseous feed through the oxidizer as inputs to predict the length of the plume and to calculate the minimum amount of heat required to ensure an invisible plume. Re-running the model as weather conditions change will lead to considerable annual savings, both in financial terms and in terms of reduced power consumption and consequent environmental impact.

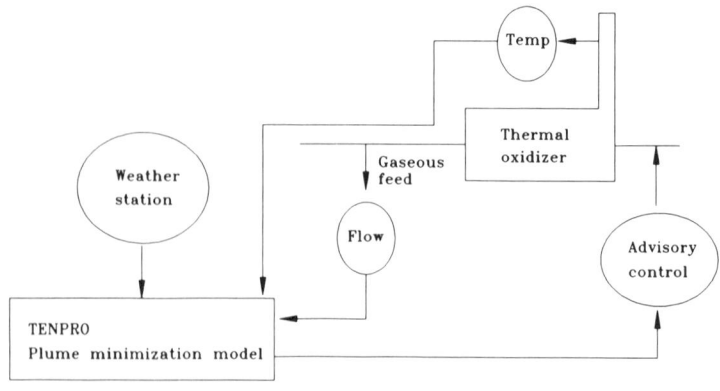

Figure 11.3   TENPRO plume suppression system.

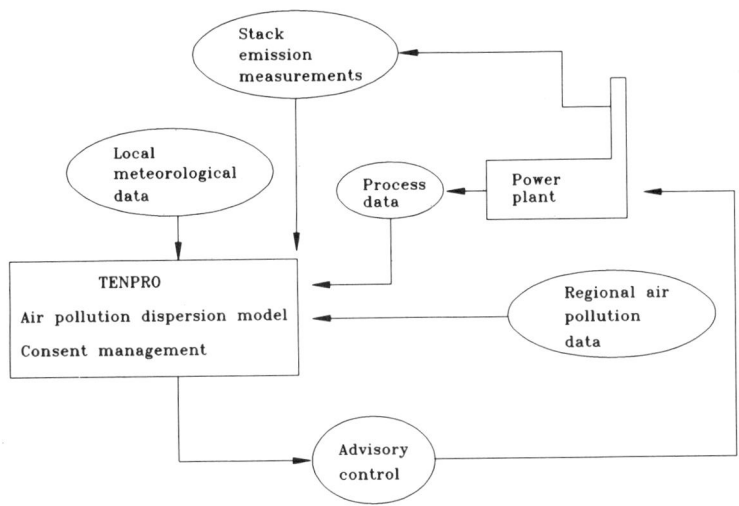

Figure 11.4   TENPRO air pollution dispersion system.

In the power sector, TENPRO provides what is being called an overall environmental management system for a fuel oil-fired power plant south of Lisbon in Portugal. An overview of the system is provided in Figure 11.4. The power plant is situated close to a major population centre and a river estuary well known for its nature reserve and tourist industry. The system will be used to:
• collect environmental and process data from a wide range of sources;
• manage, on-line, all the consents applicable to the power plant and continually monitor compliance;
• manage, report and visualize the large quantities of meteorological data collected;
• provide advisory control, enabling the anticipation of specific meteorological conditions and associated pollution episodes.

The advisory control application is geared to predicting given patterns of atmospheric dispersion due to the occurrence of specific meteorological conditions. Such episodes can lead to relatively high ground level concentrations of atmospheric pollutants, causing local property damage and breaches of national air quality standards. The dispersion model hosted by TENPRO will allow the

131

prediction of such conditions and the resultant dispersion pattern of stack pollutants, and will enable the plant operators to take the necessary internal measures to avoid extreme situations. Hence, the costs associated with these measures — from lowering power production load to switching to cleaner fuels — and any local property damage, will be minimized whilst the global environmental performance of the plant will be optimized.

12. BUSINESS RISK ASSESSMENT — A CASE STUDY

Ian Dunbar

The techniques of quantified risk assessment (QRA) have traditionally been applied to the risks posed by accidents. The assessment typically proceeds through four steps:

- identification of those accidents which may happen;
- assessment of the likelihood that an accident might happen;
- assessment of the consequences of an accident should it happen;
- judgement of the tolerability of the risks so identified and quantified.

The consequences of the accidents may be damage to:

- human health;
- the environment;
- business performance;
- any combination of these three.

This chapter proposes that the same techniques of systematic and quantified assessment of risk can be applied to the full spectrum of risks facing a business activity, and that the results provide an invaluable input to strategic business planning.

This full spectrum of risks goes well beyond just the risks from accidents. An assessment which compares a range of different types of risk within a single framework is sometimes called integrated risk assessment. When the objective of the exercise is to identify and prioritize threats to a business activity, then it can be called integrated business risk assessment.

This chapter is divided into two main sections. The first discusses the principles of integrated business risk assessment and its application to strategic business planning. The second illustrates the application of these principles, by means of a case study.

BUSINESS RISK ASSESSMENT

A *risk* associated with a given activity can be defined as an unwanted event which is a possible, but not inevitable, consequence of that activity. Unwanted outcomes of activities which definitely will happen are *costs* rather than risks. The concept of risk, therefore, has two ingredients: an undesirable *consequence*, and the *probability* (lying between zero and one) that the consequence will actually occur. In quantified risk assessment the *risk measure* is taken to be some combination of the probability and a quantified measure of consequence (for example, the number of fatalities).

THE SPECTRUM OF RISKS
There are many different types of risk associated with business activities. It is important to distinguish between a classification of risks according to their cause (risks from ...) or according to their effect (risks to ...). The spectrum of 'risks from', classified according to cause, is:
• accidents (plant malfunction, operator error, external events such as earthquake, storm, flood);
• security (malicious human action such as vandalism, arson, theft, computer hacking);
• technical (a process relied on fails to work);
• management (poor personnel management, poor project management);
• financial (movements in interest rates, loss of bank confidence, failure of flotations);
• commercial (poor sales of product, undercut by competitor);
• political (adverse action by pressure groups or politicians);
• regulation (new, tighter regulations, more stringent interpretation of existing regulations);
• company culture (attitudes of staff at different levels, especially to change);
• legal (product liability, breach of patent);
• assets (contaminated land, structural defects to buildings).

Turning to the effects of risks, the classification of 'risks to' is:
• human life (employee and public, prompt and delayed fatality);
• human health (employee and public, acute and chronic effects);
• environment (air, water, land pollution, damage to plants and animals, effects on humans);
• assets (damage to plant and buildings, contamination of land, loss of intellectual property);
• management (loss of management time, diversion of management from key functions);

- finances (cost of accidents or other damage, loss of banker/shareholder confidence);
- commercial (loss of production, sales or market share, advantage given to competitor);
- corporate image (damage to brand image, loss of supplier/retailer/customer confidence);
- industrial relations (loss of production, loss of management credibility);
- political standing (attract adverse attention of pressure groups or politicians).

For each cause and effect pairing there is a specific type of risk — for example, the risk from accidents to human health. A given risk can, of course, have multiple or knock-on consequences, and produce a longer cause-effect chain. As an example, consider an accident where the immediate consequence is damage to human health. This can in turn lead to legal liabilities which then cause financial loss to the business.

DEFINING THE SCOPE OF A RISK ASSESSMENT

The lists of causes and effects show the diversity of risks which have to be considered as part of integrated risk management. The bullet points (though not the examples given in brackets) are intended to form a complete list of risk types. Because the potential scope of a full-scale integrated risk assessment can be very large, it is important at the outset of any such exercise to define precisely what is to be included and what is to be excluded. The usefulness of having a complete list of risk types is that if the choice is made to exclude a certain range of risks from consideration, this can be done explicitly, and this exclusion can be taken into account when drawing strategic conclusions from the outcome of the study. It is far better to consider a category of risks and then decide to exclude it from further consideration than to be unaware of its existence.

The scope of a risk assessment can be defined in terms of what can ultimately be damaged by the risks. This can be:

- the business as a whole — might it cease to be profitable, or indeed cease to exist?
- a particular project undertaken by a business — a project is a one-off activity with a beginning, middle and end. Various risks might cause a project to overrun on time or budget, or might prevent the desired objective being achieved;
- a particular ongoing activity undertaken by a business — an example of an 'ongoing' activity is a manufacturing process, with raw materials coming in at one end and products and wastes going out at the other. Risks to the activity

might reduce its efficiency or cause a temporary or permanent shutdown.

Having defined the scope of the assessment in such a way, then one must define what range of initial causes is to be considered. The choice may be to look at the full range as defined above, or, after an informal assessment of the likelihood of different causes, to consider a restricted range in more detail.

This way of defining the scope of a risk assessment can be thought of as 'top-down', as it starts from the final adverse effects of the risks and seeks to determine all ways in which these effects can be caused. An alternative approach is to start from a very specific cause and then assess all possible adverse effects of the cause. An example of such a 'bottom-up' study is to ask, 'What are all the consequences of having a fire in this warehouse?'. The assessment then works its way through threats to human health and safety, threats to the environment, threats to the company's finances and so on.

This discussion has been of adverse events. It is worth noting that the same framework can be used equally well to assess the likelihood and consequences of unpredictable favourable events. This sort of assessment can help put in place a business strategy for recognizing and taking best advantage of such opportunities should they occur.

THE CASE STUDY

Due to the recession of the early 1990s, a company in the process industries was under pressure to cut production costs. The manager of a site where the production costs were higher than those on other sites investigated ways of cutting costs, and hit upon the idea of using an alternative fuel blended from used or impure solvents or other waste organic liquids. Although the use of the alternative fuel proved to be successful, it was insufficient to solve the underlying problems of the site, which were thermodynamic in nature. The company was sufficiently interested in the idea, however, to start burning the blended fuel at one of its other facilities. This was done without carrying out a full risk analysis. The technical problems were solved, and authorization to burn the blended fuel obtained. The workforce had initial reservations about the new fuel and associated handling procedures. AEA Technology helped them resolve this issue.

The company then suddenly encountered a lot of resistance from the local community, urged on by pressure groups. Its response to this was 'fire-fighting', responding to each threat as it happened with no overall plan. Realizing this was happening, the company asked AEA Technology to develop a strategic plan based on a business risk assessment. Prior to this they had used public relations consultants, who started some useful initiatives, but which were not part of an overall strategy plan. Even the early stages of the risk assessment study provided useful input to the formulation of a public relations strategy plan.

HAZARD IDENTIFICATION STRUCTURE

The aim of the assessment was to identify and prioritize the full range of hazards which could affect the burning of the alternative fuel. The requirement of completeness indicated that a structured approach to hazard identification was needed. This was developed in two stages. The first was to analyse the whole process of using the alternative fuel as a network. In this case a simple linear network was sufficient, each node of the network being one of the stages along the path from the initial generation of waste organic liquids to the final disposal of the combustion products.

The six nodes were as follows:

Waste generation

Many organizations generate the wastes which can go into the fuel, from the food industry and chemical companies through to pharmaceuticals plants and even small garages. The issue here is that the quantity and quality of the final fuel produced are largely determined by the economic climate and external pressures put on the generators.

Waste blending

The wastes arising at any one time have to be blended into a mixture suitable for use as a fuel. While the composition of the fuel can vary considerably, it is subject to tight limits on various constituents. This specification was devised by the client company, and is carefully monitored.

Transport

The blended fuel is transported by tanker to the site where it is to be burnt.

Handling

The fuel is unloaded at the client's site and piped to the facility in which it is burnt.

Burning

The fuel is burnt, ending up as deposited solid residues and airborne residues (gaseous and particulate).

137

Disposal
The solid residues are disposed of by landfill, and the airborne residues by dispersal via a stack to the atmosphere.

The burning of the alternative fuel is vulnerable to problems at any of these nodes. The second stage in the development of a hazard structure was listing the 'parameters' — the various factors which could influence or interrupt the process at the nodes. These parameters, grouped under three main headings, were as follows:

Pressure
• Workers — the workforce may be concerned about the risk of handling the fuel; they may either refuse to work or ask for unrealistic assurances or unrealistic compensation payments.
• Public — local residents, as well as any members of public, are generally concerned about waste disposal issues and may ask questions or even start a campaign of some form to stop the burning of the fuel.
• Pressure groups — some organized environmental pressure groups are against most forms of waste disposal and regularly mount campaigns against chemical companies and those in the waste disposal industry.
• Regulatory — regulators and inspectors interpret the existing legislation and police operations which involve emissions to the environment. The job of regulators involves a good deal of interpretation of regulations, and they can be open to persuasion either by the operators or by the pressure groups.
• Legislators/politicians — these people set the standards and write the legislation for companies to follow. They can be lobbied by operators, pressure groups and the general public. They have a link to the regulators. They may cause trouble for operations purely to gain party political advantage.
• Media — the media have the capability to interact with all of the above groups, to dramatize and thereby amplify the pressures (providing it makes a good story). They can scare the public and workforce, who in return will influence legislators and regulators. The pressure groups are likely to feed the media. The media do not have the attention span comparable with pressure groups and can get bored easily, but once determined to destroy a victim they will escalate the affair until a conclusion is reached.

Technical
• Engineering problems — these may limit a company's ability to meet an

objective, in this case the burning of the alternative fuel.

• Impurities — the presence of unexpected, and especially undetected, impurities in the fuel could lead to various other technical and managerial problems.

• Health effects — the toxicity of the fuel and its combustion products needs to be considered. The main issue is the ability to control the composition of the fuel. An unpredictable factor is the exposure levels allowed as tolerable by humans (inhalation or other route), which may change in the future.

• Environmental effects — this parameter is similar to health effects, and is the area where the pressure groups are particularly likely to exert pressure.

Commercial

• Fuel costs — the objective of burning alternative fuel is to reduce production costs. The financial incentive to burn the fuel can be eroded by competitors entering the market, or the suppliers and producers finding alternative routes for disposal of the wastes.

• Volume — to be successful, the alternative fuel project must be capable of responding to fluctuations, both in the demand for fuel volume and in the supply volumes of waste.

• Management costs — the management of the burning of alternative fuel brings with it additional costs, including the costs of safety requirements, management of suppliers, the Quality Assurance requirements, as well as addressing issues raised by the 'pressure' parameters.

• Legal — there are several ways in which the alternative fuel project could give rise to legal costs: fighting legal challenges in the court, challenging regulator decisions, fighting or settling compensation claims.

The resulting hazard identification structure consists of a matrix of nodes and parameters, in this case a 6×14 matrix. The network of nodes is particular to each study but, with perhaps some minor modifications, the parameter set should be applicable across a wide range of business risk assessments.

IDENTIFYING THE HAZARDS

At the beginning of the assessment the following groups of people were involved in the project:

• project managers;
• local site management of the site burning the fuel;
• people who blended wastes to make the fuel;

- public relations consultants;
- company board of directors.

They were all 'fire-fighting' on individual fronts. People from the first four of the five groups were actively involved in the assessment through a number of discussions structured around the matrix of nodes and parameters. As well as drawing out the knowledge and expertise of these people, the structured discussions helped integrate their efforts towards a single goal.

The participants in each of the discussions were taken through the matrix and asked to identify the relevant threats, consequences and safeguards for the node-parameter pairs. To make the task more manageable, each of these groups was given only a subset of the node-parameter pairs to consider. There was then a final meeting of representatives from each of the four groups which rapidly went through the whole matrix to ensure overall consistency.

In addition to identifying threats, consequences and safeguards, the participants in the meetings were asked to assign to each consequence a severity and a likelihood rating, each on a 1–9 scale. The two scales were devised in association with the client. This is particularly important for the severity scale; clients know best what outcomes worry them and how they want to order the outcomes linearly on a severity scale. The scales agreed for this study were:

Severity

1 Absolute ban on or loss of use of alternative fuel
2 Temporary ban on or loss of use of alternative fuel
3 Large damages claim against company (>£10M)
4 Medium damages claim against company (£1–10M)
5 Temporary hold-up or loss of supply of alternative fuel
6 Excessive management time and cost
7 Significant management time and cost
8 Short-term increase in management time and cost
9 Irritation

Likelihood

1 Weekly
2 Monthly
3 Quarterly
4 Semi-annually
5 Annually
6 Biennially

7   Once per five years
8   Once per ten years
9   Once per hundred years

PRESENTATION OF THE RESULTS

Each consequence identified in the discussion was plotted on a 9×9 severity-likelihood plot, giving the clients a first map of the totality of risks to the project. By superimposing their own criterion of acceptability of risk on this plot, they were able to see what this criterion implied in terms of the number of safeguards which had to be implemented, and how this number would shift if the criterion were altered.

The identification and prioritization of risks showed which of the current efforts were going in the wrong direction. Even before the assessment was complete, it was clear that the clients should not go ahead with planned expenditure in non-critical areas. The single most critical issue identified by the study had been missed by all the groups involved in the project.

While the study was going on, threats to the project from public and political pressure changed and intensified. AEA Technology liaised continually with the project managers to keep in touch with this moving target and to give them interim advice. The board of directors of the company were by this time becoming concerned with the viability of the alternative fuel project, so the board was informed of the reality of the situation. The results of the study demonstrated that the project was indeed viable, while at the same time identifying its real weak points, both in the short and long term.

At the end of the project AEA Technology was able to give a presentation to the board of directors, on behalf of the project managers. This enabled the board to see the complete picture of the direction in which the project was heading, and convinced them that it was still a valuable project to support. In addition the directors were shown how they could actively contribute towards the goals of the project. True to the predictions made in the assessment, by this time the immediate threat was from the regulators, who — under pressure from the public acting via politicians — were trying to revoke their original authorizations. AEA Technology was able to advise the client on how the regulators and policy-makers work and at what level they should approach them to initiate a constructive dialogue.

At the end of the assessment the project managers had support from their board. The project continues to use the results of the study as a predictive model of threats and opportunities. The company is now using this to predict events up to six months ahead and take appropriate measures in advance.

BENEFITS OF THE STUDY

The benefits to the client company of the risk-based approach to strategic planning can be summarized as follows:

• it brought together all the people involved in the project and started them pulling in the same (and the right) direction;

• it regained the confidence of the company's board in the project;

• it focused attention on the critical issues for the success of the project, both in the short and long term;

• it gave managers a predictive model of threats and opportunities.

CONCLUSIONS

The conclusions which can be drawn from the discussion of the general principles and from the case study are as follows:

• the techniques developed in accident risk assessment can be applied equally well to a wider range of risk factors. These techniques include the identification of hazards by means of structured discussions, the quantification of likelihood and severity (either by expert judgement, the use of statistical data or theoretical calculations), the presentation of results on a severity-likelihood plot and the comparison of the results with risk tolerability criteria;

• rational strategic business planning should use the techniques of risk assessment to identify and compare the full range of risks faced by a particular business activity. Without an overall map of risks, the management response to unexpected and untoward events is likely to be purely reactive 'fire-fighting'. Planning for future contingencies is likely to fall into the error of 'always fighting the last war'. With a map of risks, actions can be prioritized, and divided into those which need to be put in place in advance and those which can be planned and then held in readiness to meet contingencies should they arise. An integrated risk assessment gives the manager confidence that entirely unforeseen events are a great deal less likely to occur.

13. RISK ASSESSMENT FOR SPILLAGE CONTAINMENT OF HIGH EFFICACY MATERIAL — A CASE STUDY

Ian McConvey

Risk assessment is a valuable tool in helping to evaluate the necessary actions and options for containment. An environmental risk assessment was carried out on a facility making a product which has high toxicity in the aqueous environment at low concentration. Secure bunding and drainage are two important factors in avoiding environmental damage through spillage. Fairly simple mathematical models can be formulated and validated experimentally to allow evaluation of the likely concentration of any spill exiting the site or plant area. When the toxicity of the material is known, it is possible to categorize spill scenarios and quickly evaluate the effect.

BACKGROUND

Spillage or abnormal release from plants can occur by a number of different mechanisms. These spillages are usually infrequent but generally have unwelcome environmental effects. The spillages may occur outside processing plants — for example, via leakage — or within the process by malfunction of control systems. Spillages may also occur from non-process plant items — for example, tanker traffic or storage areas such as warehouses.

Spillage control by detection with instruments close to or actually on the plant can be simple and relatively inexpensive, since in this position the effluent is at maximum concentration and contains the smallest range of components to be analysed. However, if the number of plants on a site is very large, the toxicity of most of the components is quite low and the site has a common drainage system, then a common catchment system is a more effective way of providing overall spill protection.

An on-line spillage containment system is in place at the Zeneca Huddersfield site; Figure 13.1 (see pages 144 and 145) is a simplified schematic of the system. The final design includes three diversion points; the criteria used to identify an abnormal effluent at each point are as follows:

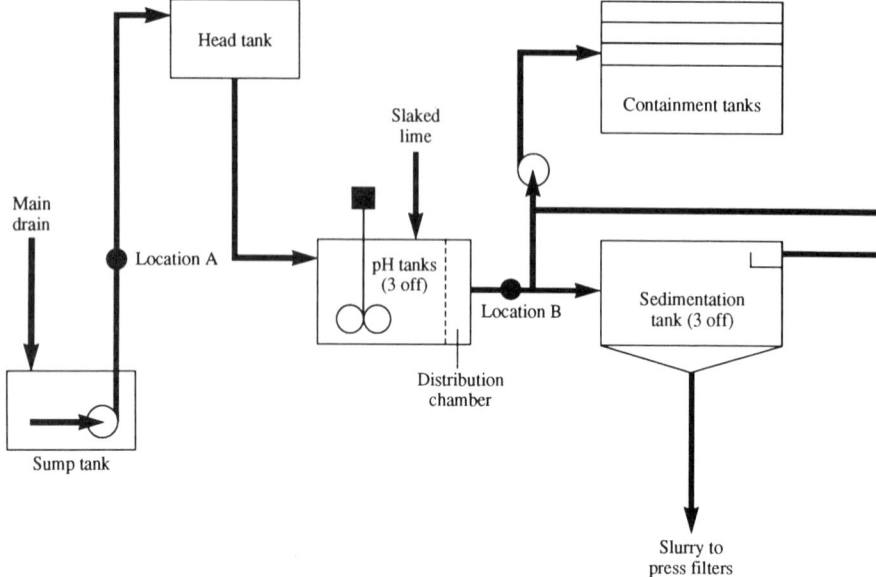

Figure 13.1   Simplified flow sheet of Huddersfield site effluent system.

Location A (immediately before pH control tanks):
(i)   Two successive abnormally high readings on two of three analysers with a reading greater than 2000 ppm.
(ii)   Presence of toxic or biologically active contaminants reported by upstream plants.
(iii) Colour contamination reported by upstream plants, effluent treatment plant operators or Yorkshire Water Service (YWS) operations. Spectrum analysers will eventually fulfil this duty.
(iv) Report of a fire giving rise to run-off water.

Location B (after pH control tanks and before sedimentation tanks)
(v)   pH too basic as measured at the exit from the pH control tanks.
(vi) Reports of abnormal discharges from upstream plants received too late to divert at Location A.
(vii) Abnormal smell observed.
(viii)Abnormal appearance observed.

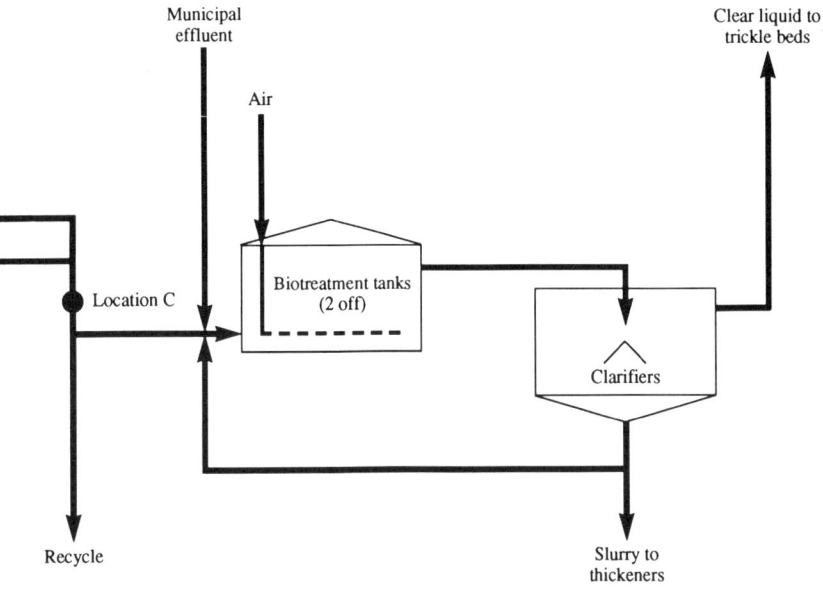

Location C (after sedimentation tanks and before outflow to YWS)
(ix) Reports of abnormal discharges from upstream plants received too late for diversion at Locations A or B.
(x) Abnormal smell observed.
(xi) Abnormal appearance (including high solids content).
(xii) High total carbon (TC) alarm on outfall analyser.

The effect of any spill passing through the system was demonstrated by carrying out tracer tests. A mathematical model of the system was written in ACSL[1], including the prediction of control valve behaviour, and this was validated against the tracer test results.

It is believed that approximately 500–1000 kg of material containing 50% by weight carbon over 1 hour would be required to give a big enough variation in the TC at the pump house to enable an incident to be detected. If plants have the potential to create incidents which would fall outside these limitations, local arrangements should be made to contain incidents. Highly toxic, low

quantity chemicals cannot be detected by the on-line instrumentation. Typical options for local spillage containment include control at source and control after spillage has occurred. Control at source can be achieved by:
- assessing risk and installing on-line detection or protective systems;
- providing local containment — for example, bunding of high risk areas.

Control after spillage can be achieved by:
- providing analysis facilities for most toxic components and diverting to site catchment facility;
- building one or more catchment interceptors locally with diversion facilities.

TOXICITY OF MATERIAL AND SPILLAGE LIMITS

The target level for the product of interest in this case study in the site aqueous effluent is 5 ppm for normal operation. The 'no effect' level is recorded by Zeneca's Brixham laboratories at less than 10 ppm. The product has an adverse effect on the respiration of the biomass in the activated sludge process, and any spill which causes the activated sludge plant to be rendered inactive could cause a major interruption to site operation. In this case, however, the trickling bed filters which are less prone to toxic shock will effect some level of treatment before outfall to the river (see Figure 13.1).

For the purposes of the product under consideration the boundaries for abnormal release could be set by using the results from the mathematical models. Spillage size may be classified as follows:

'Minor' spillage:
- duration < 0.25 hour;
- concentration < 100 ppm at source;
- mass < 1.25 kg of active ingredient.

'Significant' spillage:
- duration > 0.25 but < 1.0 hour;
- concentration > 100 ppm but < 3000 ppm at source;
- mass > 1.25 kg but < 5 kg of active ingredient.

'Major' spillage:
- duration > any time;

146

- concentration > 5000 ppm at source;
- mass > 20 kg/h of active ingredient.

Typical examples of a 'minor' spillage are vessel clean-out for maintenance or repair, pipeline leakage and surface washdown. 'Significant' spillage typically is the loss of contents of a product isolation vessel. A 'major' spillage constitutes the loss of contents of a formulation vessel or full open failure of a major reaction vessel outlet.

At the concentrations encountered, the water chemistry is such that the product is soluble, even when the pH is adjusted by the addition of calcium hydroxide and gypsum is precipitated.

LOCAL DRAINAGE

There are 17 drainage points in close proximity to the main plant and nine close to the formulations plant, not including the various surface drains. After this risk assessment a project was carried out on site in which the drain inlets for the normal process effluent were painted red whilst the clear water ones were painted blue. This alerts the operators to where any local abnormal release may be directed. Local sand bags are also provided to avoid unwanted overflow into the clear water drains.

Drainage from the south and west of the plant tends to be directed to the drain down the main avenue. Drainage from the tanker storage (on the east side of the plant) and from the formulations area enters the main drain by a circuitous route which draws effluent from other facilities. It would be difficult and disruptive to have one local catchment facility for the main and formulation plant because of the need to re-route or renew a large number of drains.

HAZARD ANALYSIS

The target for spillages or abnormal releases is obviously zero, although a higher value could be acceptable after evaluation of the criteria and frequencies have been established.

An initial target of one out-of-consent event per annum (of which, say, 10% are significant) was considered to be realistic. This would mean that direct connection of plants to drain is clearly unacceptable if the anticipated frequency is higher than this figure.

TABLE 13.1
Main plant raw materials

Material	Storage method	Comments
Raw material 1	Big bags and IBCs	Solid
Catalyst 1	Drums	Liquid
Catalyst 2	Drums	Liquid
Water treatment agents	Polytubs and drums	Liquids

TABLE 13.2
Formulation plant raw materials

Material	Storage method	Comments
Product 1	Big bags and polytubs	Solid
Caustic soda pearl	Packages	Solid
Additives (1..43)	Mostly packages	Solids and liquids

The data used for the hazard analysis has either been gathered from experience on site or using professional judgement to estimate the likelihood of any event.

Tables 13.1 and 13.2 list the main and formulation plant raw materials and their method of storage. Table 13.3 shows the large process vessels and indicates whether they are bunded or not.

BULK STORAGE

Bulk storage is treated separately because of the frequency of filling, the large quantities stored and the need to consider the appropriateness of bunding or its current status. Bunding:

• provides initial material capture in the event of a leak and this makes recovery easier;

• satisfies the regulatory authorities;

• avoids difficult or expensive retrofits if product changes and thereby improves the flexibility of the facility.

Tables 13.4 and 13.5 list the raw materials which are stored in bulk. Bunding is available on a small number of vessels.

TABLE 13.3
Large process vessels in main plant

Material	Volume, m³	Bunded?
Intermediate hold	50	Yes
Product isolation	30	Yes
Filtrates hold	30	No
Solvent separation	30	No
Raw material 4 recycle	7.5	No
Recovered solvent	10	No

TABLE 13.4
Bulk storage for main plant

Material	Volume, m³	Bunded?
Raw material 2	40	Yes
Raw material 3	40	No
Raw material 4	40	No
Raw material 5	40	No
Hydrochloric acid	60	No
Caustic soda liquor	60	No

TABLE 13.5
Bulk storage for formulation plant

Material	Volume, m³	Bunded?
Solvent 1	55	No (curbed)
Solvent 2	55	No (curbed)
Formulation 1	40	Yes
Formulation 2	10 (2 off)	Yes (half capacity)

149

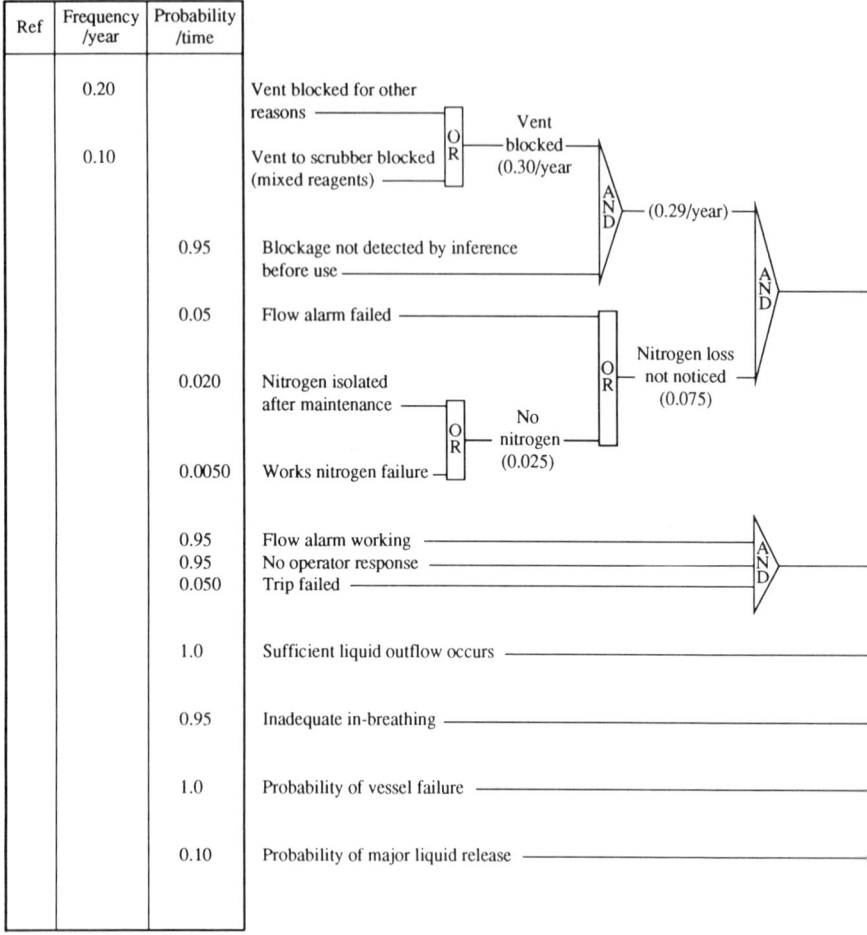

Ref	Frequency /year	Probability /time
	0.20	
	0.10	
		0.95
		0.05
		0.020
		0.0050
		0.95
		0.95
		0.050
		1.0
		0.95
		1.0
		0.10

Vent blocked for other reasons

Vent to scrubber blocked (mixed reagents)

Blockage not detected by inference before use

Flow alarm failed

Nitrogen isolated after maintenance

Works nitrogen failure

Flow alarm working
No operator response
Trip failed

Sufficient liquid outflow occurs

Inadequate in-breathing

Probability of vessel failure

Probability of major liquid release

Figure 13.2   Logic diagram for storage failure: initial case with nitrogen trip.

The accepted failure rate for metal storage vessels is generally 1×10^{-5} per annum; this should be compared against a Works standard of 1×10^{-4} for any individual item for a major environmental incident. The margin between failure by lack of mechanical integrity and the Works standard is therefore quite tight. A hazard analysis of the underpressure case for raw material 2 for loss of containment in a bunded storage is given in Figure 13.2. The frequency for this

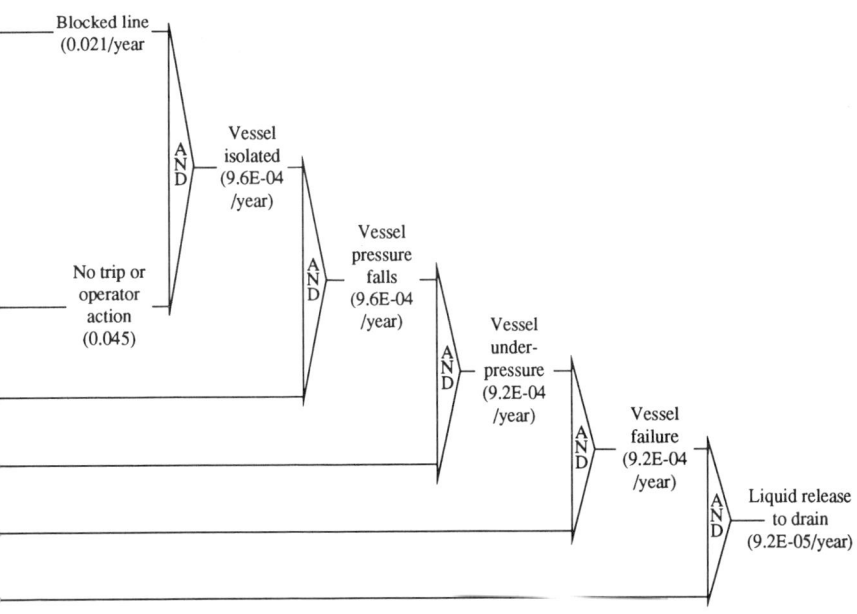

single failure mode is just within the Works standard and therefore indicates how tough it is to meet the environmental target.

RISK OF SPILLAGE

Figures 13.3 and 13.4 on pages 152–155 are logic diagrams for spillage to drain. Figure 13.3 represents the main plant and Figure 13.4 the formulations plant.

Ref	Frequency /year	Probability /time	
3.1	0.20		Abnormal release of material ————————————————
3.2		0.50	Release results in non-compliance ————————————
3.3		0.010	Not selected correctly ————————————————————
3.4		0.050	Not aware of incident ————————————————————
3.5		0.050	Aware of incident but no action ———————————————
3.6		0.050	Out to maintenance ————————————————————
3.7		0.10	Already in use ————————————————————————
3.8	0.020		Major incident ————————————————————————
3.9		0.010	Incident larger than anticipated ———————————————
3.10		0.010	Operator error ————————————————————————
3.11	22		Interceptor full ————————————————————————
3.12	2.0		Normal release ————————————————————————
3.13		0.10	Instantaneous consent exceeded ———————————————
3.14	0.10		Spurious emission from relief system ——————————
3.15		0.50	Release significatant —————————————————————
3.16	0.10		Relief valve operates ————————————————————
3.17		0.90	Release significant —————————————————————
3.18	0.10		Purge from water treatment ————————————————
3.19		0.10	Dosing to system significant —————————————————
3.20	0.10		Spillage direct to clear drain ——————————————
3.21		0.10	Spillage not noticed ————————————————————
3.22		0.50	No time to react ————————————————————————
3.23	–		Equipment failure (on clear water) —————————————
3.24		0.10	Application permits contamination ———————————
3.25	1.0		Incorrect connection to drain ————————————————
3.26		0.10	Checks not carried out ————————————————————
3.27		0.10	Mismatch causes problem —————————————————
3.28	0.10		Drum contents leaking ————————————————————
3.29		0.70	Drum not stored correctly ————————————————
3.30		0.30	Drum correctly stored ————————————————————
3.31		0.70	Leak not detected ————————————————————————
3.32		0.10	Leak significant ————————————————————————
3.33	0.10		Bulk storage or associated kit leaks ————————————
3.34		0.70	Bulk storage not bunded adequately ——————————
3.35	0.050		Major fire occurs ————————————————————————
3.36	0.10		Toxic gas release ————————————————————————
3.37	1.0		Drain leaks ————————————————————————————
3.38		0.90	Leak not detected ————————————————————————

Figure 13.3 Logic diagram for consent failure for liquid discharges: general effluent diagram with estimated failures.

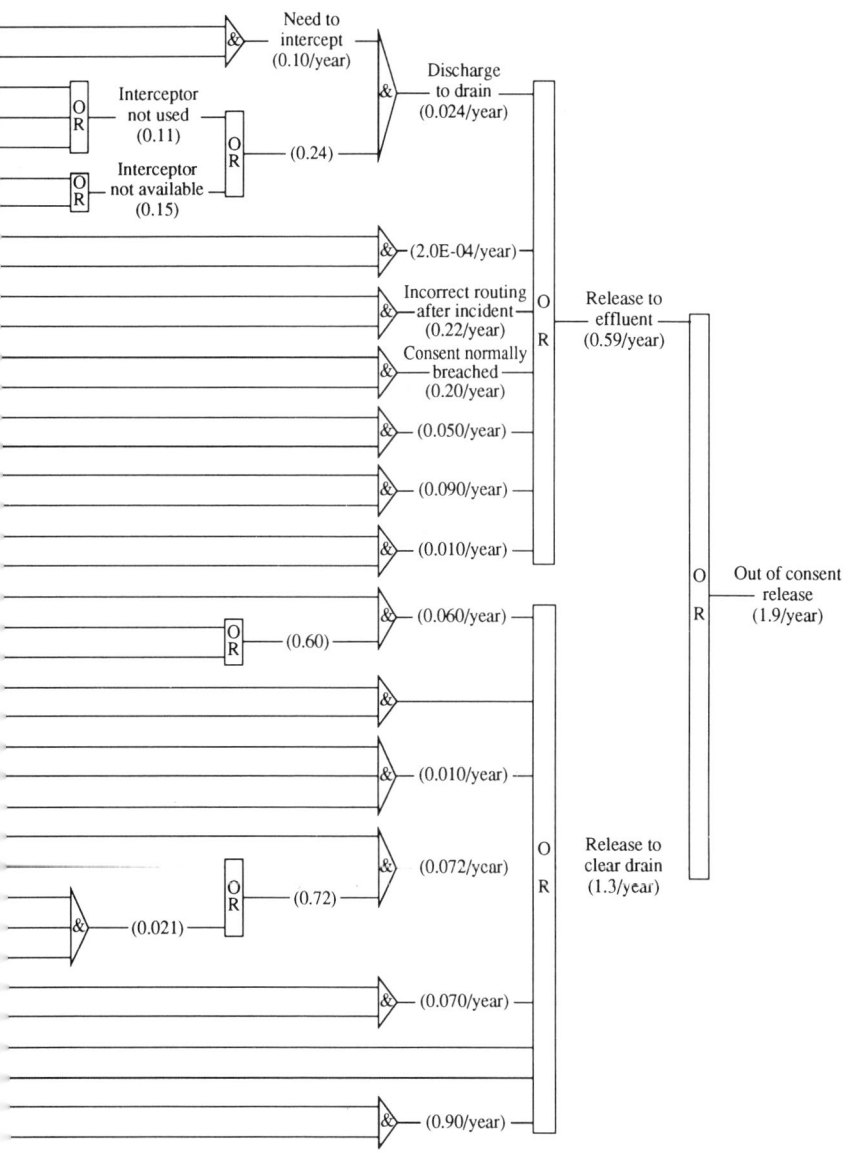

Ref	Frequency /year	Probability /time	
4.1	1.0		Abnormal release of material ————————————
4.2		0.50	Release results in non-compliance ————————————
4.3		0.010	Not selected correctly ————————————
4.4		0.050	Not aware of incident ————————————
4.5		0.050	Aware of incident but no action ————————————
4.6		0.050	Out to maintenance ————————————
4.7		0.10	Already in use ————————————
4.8	0.020		Major incident ————————————
4.9		0.010	Incident larger than anticipated ————————————
4.10		0.010	Operator error ————————————
4.11	22		Interceptor full ————————————
4.12	20		Normal release ————————————
4.13		0.10	Instantaneous consent exceeded ————————————
4.14	–		Spurious emission from relief system ————————————
4.15		0.50	Release significatant ————————————
4.16	–		Relief valve operates ————————————
4.17		0.90	Release significant ————————————
4.18	0.10		Purge from water treatment ————————————
4.19		0.10	Dosing to system significant ————————————
4.20	0.30		Spillage direct to clear drain ————————————
4.21		0.10	Spillage not noticed ————————————
4.22		0.50	No time to react ————————————
4.23	–		Equipment failure (on clear water) ————————————
4.24		0.10	Application permits contamination ————————————
4.25	1.0		Incorrect connection to drain ————————————
4.26		0.10	Checks not carried out ————————————
4.27		0.10	Mismatch causes problem ————————————
4.28	0.20		Drum contents leaking ————————————
4.29		0.70	Drum not stored correctly ————————————
4.30		0.30	Drum correctly stored ————————————
4.31		0.70	Leak not detected ————————————
4.32		0.10	Leak significant ————————————
4.33	0.10		Bulk storage or associated kit leaks ————————————
4.34		0.70	Bulk storage not bunded adequately ————————————
4.35	0.050		Major fire occurs ————————————
4.36	0.10		Toxic gas release ————————————
4.37	1.0		Drain leaks ————————————
4.38		0.90	Leak not detected ————————————

Figure 13.4   Logic diagram for consent failure for liquid discharges: general effluent diagram with estimated failures.

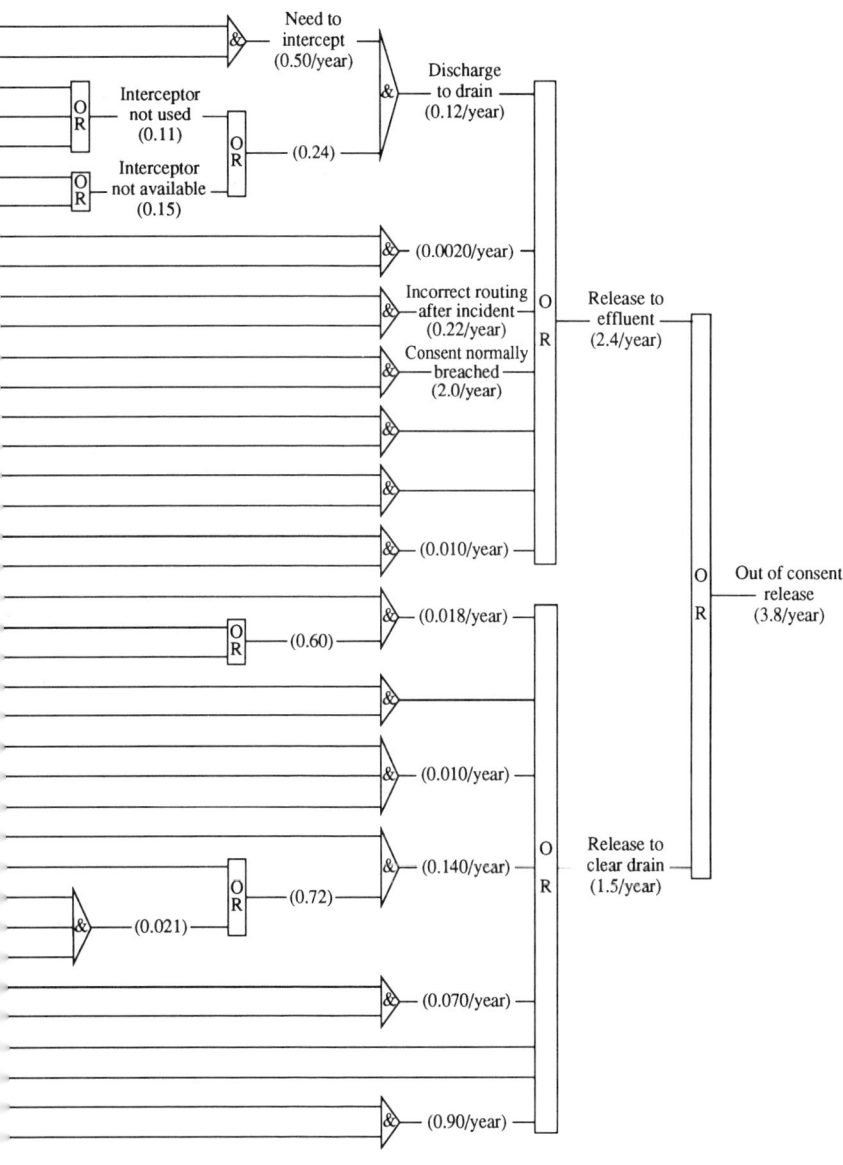

Drainage integrity and suitability is highlighted as one of the major sources of cause of spill into the environment. Clear ownership of the drains, in respect of maintenance and replacement, is important, especially on a large site. The Figures indicate that if the risk associated with the drains is removed by improvements in infrastructure then the main plant risk reduces to one out-of-consent per annum, which meets the acceptance criteria. In the formulations area the risk would drop from 3.8 out-of-consent releases per annum to 2.7. Installation of a local catchment facility system would reduce this to 1.2 out-of-consent per annum. This would be just acceptable against the criteria.

CONCLUSIONS

The case study allowed the following conclusions to be drawn:
- risk assessment is a valuable management technique which can aid the decision-making process in the environmental area between different project options and pathways;
- the validity of the data should be challenged, refined and kept under review. This process should lead to a wider use and acceptance of risk assessment;
- bunded tanks which are well managed and of the correct integrity have a very low environmental spillage risk;
- the risk assessment identifies areas into which effort should be put in order to reduce risk from spillage into the environment;
- any containment policy should include consideration of new compounds or future formulations, especially if the facility is multi-product.

RECOMMENDATIONS

- Each plant should have a clear statement of the philosophy of containment.
- Spillage containment at processing plant is improved by the implementation of on-line analysis at appropriate points, especially if it helps timely intervention in the process operation to stop any abnormal release.
- The drainage pathways, integrity and linkage should always be checked to ascertain the standard and correct routeing.
- Any hazard analysis used to manage processes should be reviewed in light of more detailed operating data or experience.

ACKNOWLEDGEMENTS
Help, advice and input from John Jackson, Steve Dempsey, Aileen Boss and Mel Holmes is acknowledged.

REFERENCES IN CHAPTER 13
1. Mitchell and Gauthier, 1975, *ACSL — Advanced Continuous Simulation Language, User Guide/Reference Manual* (USA).

14. MEASUREMENT OF PARTICULATE EMISSIONS TO AIR
Jennifer Coleman

Recent demands on industry to improve environmental performance, and the increased risk of environmental liability, have introduced the need for continuous monitoring to many industries that have no previous experience of it. Monitoring for large combustion plant and power stations has been around for some time. Indeed, some of the standard methods of calibration were developed by these industries — for example, the BCURA method developed by British Coal, which satisfies the British Standard (BS3405[1]) for particulate monitoring. This chapter is a cautionary tale for those industries which are now continuously monitoring their particulate emissions, or are considering it, either because of statutory requirements or as a means of process control. These industries have levels of particulates and types of emission which are different to combustion plant. They may have bag filters (unlike large power stations) and the process may be highly variable. Such features were found at Alcan Recycling. By following the experiences of Alcan Recycling, this chapter highlights some of the operational issues associated with particulate monitoring.

There were two types of discharges to be monitored at Alcan. The first was straightforward, whilst the second was a demonstration of the operational limitations of continuous particulate monitoring.

PLANT DESCRIPTION
Particulate emissions from two plants will be considered — the Used Beverage Can (UBC) plant and the General Products Plant (GPP). In 1991 Alcan built the Used Beverage Can (UBC) plant. This plant was designed to recycle over 55,000 tonnes per year of aluminium drinks cans and can makers' process scrap into sheet ingot for can stock production. The plant represents a significant investment for Alcan — some £28M — of which £5M was invested specifically in environmental controls. These consist of three bag filter plants ducting exhaust from every stage of the process. The first filter plant is called the 'cold system'. As the name suggests, this takes exhaust gases at near ambient temperature from an industrial hammer mill which shreds the used beverage cans

into smaller pieces of aluminium. The shredded cans then pass into a decoating system which removes all coatings from the cans. The exhaust gas from the decoater passes through an afterburner prior to discharge through a lime-injected bag filter system and out through the 'decoater stack'. The third bag filter plant exhausts the furnaces used for melting and treating the aluminium cans. This system is known as the hot system and also draws matter from the exhaust hoods above furnace doors and sidewells. The UBC feedstock is consistent and the plant runs continuously. Each bag filter system is separate and exhausts to its own stack. Each of the three stacks, therefore, carries a specific type of particle at a velocity which is largely constant. From commissioning, each stack was continuously monitored for particulates using optical meters.

Monitoring particulates on the General Products Plant (GPP) involved retrofitting, a step faced by many other industries. The GPP has one main stack taking exhaust from two systems. The first is a lime-injected bag filter plant serving two melting furnaces. The second is a common exhaust from three melting furnaces and two holding furnaces. A variety of scrap and fuels are processed in these furnaces. When monitoring was introduced there was no fume abatement in the second system. The furnaces act largely independently so there

Alcan Recycling, Warrington

is no consistent process profile. As a result, stack emissions varied from next to nothing when there was no furnace activity, to dark smoke under certain furnace process conditions.

THE NEED FOR PARTICULATE MONITORING
Alcan uses particulate monitoring to fulfil regulatory requirements, to monitor bag performance and to control visible emissions.

FULFILLING A REGULATORY REQUIREMENT
In the newer UBC plant, continuous particulate monitoring was in the original plant design and accepted as part of initial authorization to operate. Both the UBC and GPP emission data from continuous monitoring are used to calculate monthly emission values, which are reported to HMIP.

MONITORING BAG PERFORMANCE
At Alcan Recycling information on emissions has been utilized in more ways than just fulfilling a clear regulatory requirement.

Bag filters deteriorate over time — indeed, many manufacturers talk about 'expected bag life'. This deterioration can be accelerated by a number of factors, including acid gas attack and high exhaust temperatures. Emission trends can be used to indicate bag performance. An overall upward trend can mean that the bags are nearing the end of their useful life. This can be confirmed by laboratory analysis of a number of bags for permeability, density, thickness and bursting strength. Bag manufacturers may offer this service free of charge. Operators should keep a log of which bags are removed because the mistake may be made of submitting new bags for analysis, which are obviously not representative of the remaining bags. By pro-actively monitoring bag performance the risk of exceeding emission limits is minimized.

A regular emission spike that occurs during a cleaning cycle indicates catastrophic deterioration in one or several bags. The bag(s) can be identified by leak testing the bag filter plant with lime or preferably fluorescent powder. Visual inspection for fresh dust on the clean side of the filter reveals which bag(s) needs replacing. By monitoring when the peak occurs, it is often possible to isolate the leak to one module of the bag filter plant. This saves time on larger plant where there are many modules, as only the suspect module needs to be tested for leaks.

VISIBLE EMISSIONS

Alcan Recycling is sited in a built-up area with a number of other industrial stacks nearby. For many sites close to a residential area, smoke can be a source of public concern. Alcan has used continuous monitoring as a tool for proving that a particular cause of complaint was not from an Alcan stack. If used wisely it can also be a good tool for reacting to any emissions. There are a number of different alarm systems in use at Alcan that inform plant operators of emission levels. For example, large LED displays of dust levels are easily visible in the foundry. Alarms sound and a message of high dust levels is printed. In addition to the dust level system, audible and visual alarms are activated should a fume arrestment plant go into bypass. By these means any rise in dust levels is picked up quickly by those people operating the plant, who will take the necessary steps to limit the event. This minimizes the emission, reduces the number of public complaints and improves relations with the neighbours and the regulatory authorities.

INSTALLATION

Continuous dust monitors were installed in the exit ducts from both systems in the GPP. External contractors carried out extractive sampling in order to calibrate the meters. The meter on the bag filter plant worked successfully, recording readings around 3 or 4 mg/m^3.

This level of emission was expected from a plant with a bag filtration system and was in line with the experience in the UBC plant. The meter on the un-arrested plant, however, told a different story. A meter would record maximum levels of 999 mg/m^3 which did not correspond to the actual emission — for example, a reading off scale would be recorded when only light grey smoke was visible from the stack. At first the positioning of the meter was questioned. Sensor heads had been placed close to the main fan at a point that was easily accessible, but where the flow was extremely turbulent. This point had not been picked up by the external contractors brought in to take iso-kinetic samples for calibration purposes.

Consequently, alternative contractors took samples and advised on the appropriate siting of the monitors. Detailed advice is available in HMIP Guidance M2[2] on the positioning of stack monitors. Briefly, monitors should be sited away from obstructions in the duct work that are likely to cause turbulence — for example, bends, fans, manholes and changes in duct diameter.

Siting monitors can be a costly activity as it may involve the construction of permanent platforms or, at least, temporary scaffolding to allow access for maintenance and regular sampling for calibration. For this reason permanent platforms large enough for sampling apparatus and teams were installed. BS3405[1] gives information on the siting of ports for monitoring and sampling. It may be the case that existing duct work does not have runs that are long or straight enough to be the required distance from the nearest obstruction. Ports should be sited in an area where the flow is approximately laminar. One should remember to take into account access manholes and existing ports or holes. If a port is an addition to many others, the structural integrity of the duct or chimney should be considered. Adding more holes may have the effect of perforating the top of the chimney!

CALIBRATION

EXTRACTIVE SAMPLING
In order for any continuous dust meter to work quantitatively it must be accurately calibrated against an iso-kinetic sampling technique such as BS3405. The accuracy of the continuous meter is therefore highly dependent on achieving an accurate calibration.

Thus, in addition to ports for continuous monitoring, equipment sampling ports are required for the iso-kinetic extractive sampling. To comply with BS3405 a number of ports should be fitted so that samples can be taken at positions across the plane of the stack, effectively splitting the duct into portions of equal size. In this way a representative sample of the exhaust gas can be obtained. The specification for round and square ducts is quite different. Advice should be sought before fitting ports; it is available from the British Standard but, in Alcan's case, the local pollution inspector was more than willing to help.

As well as dividing up the stack appropriately, the dust must be collected iso-kinetically — that is, at the same velocity as the exhaust gases. But how can the plant operator be sure that the external contractors are taking the correct steps to obtain a representative sample without being there throughout the sampling period? Alcan again found that the local inspector was able to offer a handy hint; the company now asks sampling teams to provide the Pitot readings as a check.

THE EFFECT OF PARTICLE SIZE

Having re-sited the instrument in the GPP stack, the problem of off-scale readings persisted. It became apparent that the monitors could not cope with the variability of the process.

In an attempt to overcome this problem, Alcan tried a two-point calibration. Dust sampling must be carried out over a period of time long enough to obtain a measurable quantity of dust. The sample taken must be representative of the process being measured, so the process must be held constant for the sampling period. Low dust levels are easy to establish; this is when there is no activity in the furnaces. The upper point was needed to represent the emission that occurred when all the furnaces were being processed. These periods were limited and proved not to be long enough for sampling purposes. The closest approximation that would give adequate sampling time was obtained by the generation of smoke from inefficiently lit burners.

Despite this attempt at a two-point calibration, spurious results were still recorded. It was found that, whilst monitors responded linearly to an increase in dust concentration, it was only to dust of the same type. The smoke produced from a burner is a different material with a different density and size to that produced when molten aluminium is being processed.

The problem was only resolved satisfactorily when a bag filter plant was installed to treat and abate the remaining furnaces, resulting in an emission of largely constant particle size and concentration. There are, however, occasions when a filter plant goes onto bypass and the issue of different materials arises again.

Both optical and particle impingement meters have been trialed on site. Impingement meters were found have an improved response to the small variations in emissions that occur from the filter plants in the particular arrangement of ducts. Neither method, however, has proved to give reliable data for all sizes and types of emission on all occasions.

MAINTENANCE AND INSTALLATION

Like all equipment, dust monitors require maintenance. Instruments may have built-in check facilities that indicate a fault. Alcan found that the build-up of dirt on the sampling heads of light and impingement monitors has a significant effect. The frequency with which probes or sensors need to be cleaned varies depending on the constituents of the exhaust gases. In Alcan's experience this is one of the first ports of call when questionable data is being produced. The positioning of sensing heads some fifty or more metres away from the meters has led to long lengths of connecting wire. The siting of these wires has been another cause of spurious results, due to the effects of electrical interference.

BACKUP

Until the company is confident that continuous monitors do capture every type of emission, a video camera monitors the stacks at all times. The camera also gives a final check to ensure that all smoke emissions are recorded. Should the security guard see a smoke emission the operators are informed so that corrective action can be taken immediately. This may seem like excessive monitoring, but for a highly visible recycling company it is money well spent to control the process and maintain good relations with the public and regulatory authorities. It is also an indication of the growing importance of environmental management to the industry of today.

REPORTING

Many companies, like Alcan Recycling, have been reporting emission data either to regulatory authorities or internally for many years. It is a likely requirement of IPC. Those companies already doing it will know that it is a time-consuming operation. Alcan has found that for the data to be of any use it should be inspected daily. For example, a burst filter bag can have rapid dramatic effects on emission levels and should be isolated as soon as possible.

To aid the downloading of data, modem connections have been or are being fitted to the meters. This significantly reduces the time taken in obtaining the data, which can then be spent more productively analysing data trends and establishing the cause of an emission.

Presentation of data is another point to consider. Over what period should the emission be averaged? Will smoothing of the data help to diagnose slowly deteriorating bags? This may be the factor that distinguishes one type of monitor from another. Some provide a list of average emission figures without allowing flexibility in the way the average is taken. Others automatically graph any portion of the downloaded data over any averaging period. These data can be investigated at any time and manipulated to show overall trends or incidents in more detail. This is a very useful feature if one is trying to isolate an incident that has not affected the emissions recorded over the longer averaging period.

CONCLUSIONS

There are many different monitoring devices available, each with its own particular strengths and drawbacks. They fall broadly into two categories — optical meters and particle impingement meters. Alcan Recycling has tried both types and neither has proved to be ideal for every situation. The company's experiences highlight some of the pitfalls common to monitoring equipment and some of the other issues worth considering when monitoring particulate emissions.

REFERENCES IN CHAPTER 14

1. BSI, *British Standard 3405: 1983 (1989), Method for Measurement of Particulate Emissions including Grit and Dust.*
2. HMIP, November 1993, *Technical Guidance Note (Monitoring) M2: Monitoring Emissions of Pollutants at Source* (HMSO).

15. RISK ASSESSMENT — A USEFUL TOOL FOR THE ENVIRONMENTAL MANAGEMENT PROCESS

Wolfgang Ihme, Marcus Ford and Richard Andrews

Environmental management systems (EMS) are designed to manage and control the activities of industrial sites or organizations in order to prevent egress of materials to the wider environment. Most sites, however, have a history of emissions — ranging from single incidents to prolonged releases — which have caused some form of impact to the soil, water or air environment, and perhaps the species occupying such areas, including man. Normally the highest degree of attention is paid to potential exposure of human beings and related health risks. When impacts are obvious — for example, fugitive dust emissions causing nuisance to local residents or movement of contaminated run-off to a receiving stream — abatement measures can usually be implemented forthwith. However, for soil and groundwater contamination the problem can go unnoticed for years or even decades, something that often compounds the management problem.

Soil and groundwater contamination issues pose another dilemma to a site environmental manager. The impact that any contamination may have is very much governed by site-specific issues and controls. Once released to the soil environment, hazardous chemicals may not pose a significant risk to either man, fauna or flora on site. On the other hand, specific contaminant physico-chemical properties and specific exposure situations may mean that not only are employees affected, but released chemicals may also be spread by air or migrate in an aquifer. Then, more distant receptor groups may be exposed. Indeed, it may be necessary to consider the aquifer itself as a receptor, if the UK National Rivers Authority (or equivalent regulatory body) considers it to be a resource worth protecting. Given such local controls it becomes critical to answer the question 'how clean is clean?' both for an operating site and for consideration of other possible land use.

It would be easy if this question could be answered by applying generic criteria such as the Interdepartmental Committee on the Redevelopment of Contaminated Land (ICRCL) values or those given by the Dutch Health and Environmental Agency (RIVM), known as the Dutch List. Major limitations are caused by the lack of these values or by their very limited site-specific character.

167

While the current ICRCL values already consider the end use of a site to some degree, the Dutch List quite deliberately refers to multifunctional land use. This is obviously contrary to the UK Government's 'suitable for use' approach that is preferred in the UK for control and treatment of contamination[1].

These problems can effectively be addressed by identification and evaluation of actual risks posed by contaminated land. Risk assessments can be undertaken at two levels — qualitative or quantitative. The level selected depends upon contamination levels, site setting, intended land use and the level of detail and confidence required. The assessment of true liability and the formulation of procedures for the control and management of these risks can be undertaken during the risk management process.

QUANTITATIVE RISK ASSESSMENT

The risk assessment process can be defined as a combination of methodologies all aiming to evaluate the likelihood of an adverse effect on humans, domestic animals, wildlife or ecological systems produced by a specific level of exposure to a chemical or physical agent[2]. According to the United States Environmental Protection Agency (US EPA)[3] this approach may be regarded as a four-stage system comprising:

- data collection and evaluation;
- exposure assessment;
- toxicity assessment;
- risk characterization.

During *data collection and evaluation* all available data are compiled and evaluated with respect to their application in the later process. This process includes a Quality Assurance (QA)/Quality Control (QC) task. Based on a qualitative evaluation of physico-chemical properties, toxicological data and site-specific information, a conceptual site model (CSM) is designed. It qualifies potential exposure pathways with respect to current or future land use of the specific site. As part of this exercise, the number of compounds to be considered during the next steps of the risk assessment may often be reduced to so-called 'priority contaminants'. Such confinement helps to focus on the most significant compounds (on a site-specific basis) and to keep the risk assessment process within a realistic time and budget frame.

Major data gaps may be discovered during the data collection and evaluation stage. It may also appear that too many data — and, in particular, inappropriate or unnecessary data — have been collected during previous investigations. It may prove superfluous to take soil samples for analysis from a soil layer several metres down, but highly important to know the concentration in the top soil layer. For instance, if a non-mobile compound (that is, non-volatile and of low solubility) only poses a risk to humans by ingestion or dermal contact, the risk assessor must focus on the upper and therefore accessible soil layer because exposure to deeper soil layers is rather unlikely or even impossible.

Spatial statistical techniques (geostatistics) may be used during the data collection step to reduce uncertainty in the risk assessment process significantly at the sampling site investigation stage. They may also provide more meaningful and realistic averages and 95% upper confidence limits for the parameters required in exposure calculations. Geostatistics may be used to design efficient and focused sampling plans for a given spatial variable (for example, the concentration of benzene in soil) to a particular level of statistical confidence, typically the 95% confidence level. This approach improves the general site characterization process, understanding of the nature, magnitude and extent of contamination, and assures relevant authorities or other third parties of the thoroughness of an investigation.

For typical contamination investigations, samples are targeted in areas of a site that are most likely to be contaminated. This 'common sense' approach clearly biases the data and means that classical statistical approaches are invalid and cannot be used to assess the data. To do so would give a biased average for the site as a whole and is likely to suggest that the general level of contamination is much higher than in reality. Using such biased data in a risk assessment may then result in the estimation of higher levels of apparent risk and a recommendation for remediation that may, in fact, not be required. Geostatistics may be used to 'unbias' the data and provide realistic estimates of the actual distribution of concentrations across the site together with an estimation of the associated error. Averaging over the area of the site, or a previously designated area of a site, is typically undertaken in risk assessments since it provides a more representative assessment of the general level of exposure to a compound than classical statistical techniques, which are unable to take account of the spatial distribution of different levels of contamination.

During the next stage, the *exposure assessment*, all site-specific information on potential exposure is compiled. A detailed evaluation of the exposure pathways that link a chemical release point with a potential receptor population is undertaken. Quantitative data on several exposure parameters are required.

Such parameters comprise age-dependent or behaviour-specific data (ie, how often and how long and to what degree may the receptor be exposed: for example, intake rates for soil, dust, air or food), site-specific data (for example, climate conditions) as well as compound-specific data (for example, soil or groundwater concentrations or partitioning coefficients for soil matrix/soil water). Fate and transport models often have to be used to relate local soil, groundwater or air concentrations to distant exposure conditions, which cause a significant increase in the exposure data set.

The *toxicity assessment* is designed to evaluate the toxicity of each compound or compound group. It is completely independent of any site-specific data (except the list of priority compounds considered). Various qualitative and quantitative information on human and non-human toxicity is collated and evaluated in order to characterize a compound by a single item or a few items of toxicological data.

One important feature which has an impact on the way in which calculations are carried out is whether the compound in question is carcinogenic or non-carcinogenic. It is commonly assumed that carcinogens (that is, promoters) do not have a threshold below which no risk of cancer exists — that is, there is only zero risk if there is no exposure. On the other hand, non-carcinogenic compounds are regarded to show a threshold below which there is no adverse effect. These considerations are reflected by two groups of toxicological criteria derived during the toxicological assessment. One is comprised of potency factors and the other of maximal acceptable exposures. A potency factor quantifies the incremental lifelong risk of cancer caused by a lifelong exposure to 1 mg/(kg d) of the respective chemical. A maximal acceptable exposure — often given in mg/(kg d) — gives the dose below which no adverse effect is likely to be observed, even after lifelong exposure. The Reference Dose (RfD) derived by the American Environmental Protection Agency, or the Acceptable Daily Intake (ADI) derived by the World Health Organization, may serve as examples for maximal tolerable exposure criteria.

Finally, during *risk characterization*, information resulting from the preceding stages is combined, giving a risk assessment that is site-, use- and compound-specific. The evaluation is health-protective, driven by the application of conservative assumptions and parameters. This ensures that even under unfavourable conditions receptors are still sufficiently protected. A risk characterization can either include the quantification of exposure and risk to potentially exposed receptors or the derivation of health-protective concentrations in soil, groundwater or any other medium. Such concentrations may guide the development of a remedial goal.

The theoretical fundamentals of both approaches will now be given, together with case studies to illustrate them.

EXPOSURE RISK CALCULATION

The exposure of humans to any compound in the environment can be calculated by the following standard equation, which determines the dose of a compound that enters the body. This equation represents the basic algorithm in various exposure models[3,4,5].

$$E = \frac{IR \ T \ C}{BW \ AT} \tag{1}$$

where:

E = exposure to a compound via the pathway concerned, mg/(kg d);

IR = intake rate per day of the medium concerned (for example, food, soil, air), kg/d;

T = exposure time, d;

C = concentration of the compound in the medium taken up, mg/kg;

AT = averaging time (for example, confined period or lifetime), d;

BW = body weight, kg.

While equation (1) gives the contacted dose (as it reaches the body barriers), the absorbed dose can be calculated by multiplication with the appropriate absorption fraction that applies to the particular pathway. The absorbed dose reflects the amount of contaminant that actually enters the blood stream.

Such exposure estimations may be carried out for different periods of life — for example, childhood or adulthood. Then lifelong exposure may be calculated as:

$$E = {}^{15}\!/_{70} \, E(child) + {}^{55}\!/_{70} \, E(adult) \tag{2}$$

Here the lifelong exposure refers to an average life expectation of 70 years and defines 'childhood' as a period of 15 years and 'adulthood' as a period of 55 years.

Equations (1) or (2) may be used to calculate exposure for different areas of a contaminated site, different activities of receptors, and all relevant exposure pathways in agreement with the current or planned land use.

The *carcinogenic risk* is calculated by using a potency factor according to:

$$CR = PF \cdot E \tag{3}$$

where:

CR = incremental lifelong risk of cancer caused by the lifelong exposure E, cases of cancer among exposed persons;

PF = potency factor, $(mg/(kg\ d))^{-1}$;

E = lifelong exposure to a compound via the regarded pathway(s), mg/(kg d).

Cancer risks caused via different exposure routes and by several compounds are summed.

Non-carcinogenic effects are evaluated by the following equation:

$$HI = \frac{E}{MAE} \tag{4}$$

where:

HI = hazard index, giving the relation between estimated exposure on site and maximum acceptable exposure, dimensionless;

E = exposure to a compound via the pathway concerned, mg/(kg d);

MAE = maximum acceptable exposure (for example, RfD or ADI), mg/(kg d).

Depending on the compound and the type of MAE used to characterize its toxicity, it may be necessary to weight the exposure E over lifetime or a shorter period. Similarly the type of MAE drives whether exposures by specific pathways are to be summed and how this is performed.

DEVELOPMENT OF SOIL VALUES (REMEDIAL GOALS)

In general, the protection of a human receptor is regarded as of prime importance when a specific remedial goal for soil or groundwater is derived. Health-protective soil concentrations (termed 'soil values') ensure that no adverse effect is likely to be caused, even under unfavourable conditions. Often such soil values can be applied directly as remedial goals (if, for example, ecotoxicological considerations are not of concern).

In principle the development of soil values follows a backward calculation of the exposure risk calculations presented above. This is demonstrated for exposure via soil ingestion below. For carcinogenic effects, cancer risk can be described (based on equations (1) and (3)) by:

$$CR = \frac{IR\ T\ C}{BW\ AT}\ PF \qquad (5)$$

If the variable cancer risk, CR, is replaced by an acceptable cancer risk, C_a, and it is assumed that C represents a soil concentration and IR the ingestion rate of soil, the equation can be solved for this soil concentration:

$$C = \frac{CR_a\ BW\ AT}{IR\ T\ PF} \qquad (6)$$

The soil concentration equals the site- and land-use specific soil value. An 'acceptable cancer risk' is often taken to be in the range of one case per 1,000,000 to one case of cancer per 10,000 theoretically exposed persons (that is, to 1×10^{-6} to 10^{-4}). In the case studies described below, a value of 1×10^{-5} was selected.

For non-carcinogenic effects the soil value is calculated in a similar way:

$$C = \frac{HI\ BW\ AT\ MAE}{IR\ T} \qquad (7)$$

Again the simplified example addresses soil ingestion and IR refers to the soil ingestion rate. When exposure to only one compound by this pathway is to be assessed the hazard index may be taken as unity.

Both equations (6) and (7) become much more complicated if exposure via several pathways is to be considered and/or more than one compound poses a potential risk to the human receptor.

CASE STUDIES

This section provides simplified examples of the application of the risk assessment approach. It presents an exposure and risk characterization for the drinking water pathway at a site in The Netherlands and the derivation of soil values (remedial goals) for an industrial setting in the UK where soil/dust ingestion, dermal contact with soil/dust as well as inhalation of suspended particles were considered.

CASE STUDY 1 — CALCULATION OF EXPOSURE RISK
In The Netherlands clean-up of industrial facilities is often governed by the objective to achieve a multifunctional end use for the site. This entails high emphasis on groundwater as a potential source for human drinking water. This case study looks at a coating facility where contamination of soil and groundwater by benzene, toluene, ethylbenzene and xylene (BTEX) had occurred. Methyl isobutyl ketone (MIBK), mineral oil products and heavy metals had been discovered in concentrations exceeding the (old) Dutch B- and C-values (which have since been superceded).

In order to evaluate the risk caused by exposure to groundwater, the following scenario was set up. It was assumed that the (future) owner of the site would install a well to abstract water from the first main deep aquifer below the site. Based on near surface perched (shallow, unconfined) water concentration, the water concentration in the first aquifer was modelled (using a deterministic 'mixing cell' solution to the advection-dispersion equation for solute transport in groundwater). Exposure and risk could then be calculated for consumption of such water.

Table 15.1 lists concentrations and results for some of the major contaminants in the perched water including benzene, xylene, MIBK and mineral

TABLE 15.1
Exposure risk due to consumption of water abstracted from the first aquifer below a site in The Netherlands

	Benzene	Xylene	MIBK	Mineral oil[1]
Perched water conc., $\mu g/l$	20	350	500	1250
First aquifer water conc., $\mu g/l$	0.05	1.0	1.5	3.5
Exposure by water, mg/(kg d)	1.4×10^{-6}	3.6×10^{-5}	5.4×10^{-5}	1.3×10^{-4}
Cancer risk, cases per 1 million	0.04	–	–	–
Hazard index, dimensionless	–	$\ll 0.001$	0.001	0.002

MIBK = methyl isobutyl ketone
[1] n-hexane is taken as a surrogate compound for mineral oil (excluding BTEX and polycyclic aromatic hydrocarbons (PAHs))

oil products. The exposure calculation for benzene (being a carcinogen) was based on the assumption of lifelong exposure — that is, a daily intake rate of 2 litres of water from the private well was assumed, which is considered to be highly conservative. In a pragmatic approach the body weight of an adult was taken as the lifetime average (see Table 15.2, page 176, Scenario A1). Opposed to this, for the other, non-carcinogenic compounds, children were considered, representing the potentially most sensitive exposed receptor group. Therefore only a 1 litre/day consumption of groundwater was assumed (Scenario A1). These and the other toxicological parameters used are given in Tables 15.2 and 15.3 respectively, pages 176 and 177).

The results indicate that, even under such conservative and unlikely assumptions as presented above, lifelong exposure to benzene in drinking water abstracted from the first aquifer below this site would only cause an incremental lifelong cancer risk of 0.04 cases per 1 million (or 4 cases per 100 million theoretically exposed persons). The non-carcinogenic effects of xylene, MIBK and mineral oil were also negligible. For this assessment n-hexane was selected as surrogate compound for non-carcinogenic mineral oil product compounds. The hazard index reached a maximum value of 0.002 — that is, only 0.2% of the maximum acceptable exposure to mineral oil on this pathway. For xylene and MIBK the risk was even less. The results also showed that, due to further degradation and attenuation and adsorption processes in the aquifer, no threat was likely to exist to the water quality of a public water supply located 2 kilometres away.

CASE STUDY 2 — DEVELOPMENT OF SOIL VALUES (REMEDIAL GOALS)

The question of what soil concentration is acceptable for a proposed land use often arises. Such soil concentration should not allow any exposure that exceeds a toxicologically tolerable intake of a specific contaminant. If the future end use of a site is of an industrial nature, it may be expected that the soil concentration that does not pose any significant health risk to the potential receptor (that is, workers) will be significantly higher than for other types of end use — such as residential or recreational. With industrial end use, many exposure pathways can be excluded from consideration at the very beginning (for example, intake of home-grown foodstuffs).

This case study describes three typical exposure pathways often encountered on an industrial site. An area in the south of the UK was redeveloped and a potential buyer was expected to use the site for new industrial buildings.

TABLE 15.2
Age- and site-specific parameters

| | Scenario* | | | | |
| | A1 | A2 | B1 | B2 | B3 |
Parameters	Child	Lifelong	Adult	Adult	Adult
Body weight, kg	28	72	72	72	72
Drinking water consumption					
Daily water consumption, l/d	1	2	–	–	–
Exposure frequency, d/y	365	365	–	–	–
Exposure duration, y	15	70	–	–	–
Soil/dust ingestion					
Daily soil/dust ingestion rate, mg/d	–	–	100	100	100
Soil/dust ingestion in working day, mg/8h	–	–	50	50	50
Proportion of outdoor soil in indoor dust	–	–	0.3	–	–
Time fraction when exposure is possible			–	0.5	1.0
due to climate, dimensionless	–	–			
Exposure frequency, d/y	–	–	260	48	40
Exposure duration, y	–	–	15	5	5
Dermal contact with soil					
Soil/dust adherence, mg/(cm^2 d)	–	–	1.0	1.0	1.0
Exposed skin surface, cm^2	–	–	1688	5318	12,631
Proportion of outdoor soil in indoor dust	–	–	0.3	–	–
Time fraction when exposure is possible					
due to climate, dimensionless	–	–	–	0.5	1.0
Exposure frequency, d/y	–	–	260	48	40
Exposure duration, y	–	–	15	5	5
Inhalation of suspended particles					
Daily breathing rate, m^3/d	–	–	–	–	28
Breathing rate in working day, m^3/8h	–	–	–	–	9.3
Proportion soil/air conc., kg/m^3	–	–	–	–	2.4×10^{-7}
Exposure frequency, d/y	–	–	–	–	40
Exposure duration, y	–	–	–	–	5

* Scenario A1 = residents (children) — drinking water consumption
  Scenario A2 = residents (lifelong) — drinking water consumption
  Scenario B1 = workers — indoor exposure
  Scenario B2 = gardeners
  Scenario B3 = workers performing intrusive work in the ground

TABLE 15.3
Toxicological criteria

Compound	Potency factor[1], $(mg/(kg\ d))^{-1}$ Oral	Inhalative	Reference dose[1], $(mg/(kg\ d))^{-1}$ Oral	Inhalative	PTWI[2], $mg/(kg\ d)$
Arsenic	1.8	1.5×10^1	–	–	–
Cadmium	–	6.3	5.0×10^{-4}	–	–
Copper	–	–	3.7×10^{-2}	–	–
Lead	–	–	–	–	7.14×10^{-3}
Zinc	–	–	3.0×10^{-1}	–	–
Benzene	2.9×10^{-2}	2.9×10^{-2}	–	–	–
Xylene	–	–	2.0	2.0×10^{-1}	–
MIBK	–	–	5.0×10^{-2}	2.3×10^{-2}	–
Mineral oil[3]	–	–	6.0×10^{-2}	5.7×10^{-2}	–
PAHs[4]	7.3×10^1 See [5]	3.7×10^2 See [6]	–	–	–

PAHs = polycyclic aromatic hydrocarbons
PTWI = provisional tolerable weekly intake, transformed to a daily basis
[1] derived by US EPA according to Smuker (Reference 6)
[2] according to WHO (Reference 7)
[3] n-hexane is taken as a surrogate compound for mineral oil (excluding BTEX and PAH)
[4] the potency factors refer to benzo(a)pyrene as an indicator substance but evaluate the risk from all PAHs
[5] the oral risk is approximately ten times higher than from benzo(a)pyrene alone according to Slooff et al (Reference 8)
[6] the inhalative risk of all PAHs is approximately 50 times higher than from benzo(a)pyrene alone according to Grimmer et al (Reference 9)

As a result of a long industrial history, the site soils were heavily contaminated with various heavy metals including arsenic, polycyclic aromatic hydrocarbons (PAHs) and mineral oil products. On the other hand, investigations of hydro-geological conditions suggested that groundwater contamination was unlikely to impact a wider environment. This was due to the general poor water quality, the low mobility of the site-specific contamination and the absence of any water supply in the vicinity which could potentially be affected.

Due to the designated industrial end use of the area, it was planned to cover most of the site with buildings, car parks and hard standing areas. Never-theless the vendor wanted to make sure that there were no health risks to his workers or other persons caused by remaining site contaminants. For this reason three scenarios were chosen:

- Scenario B1 — workers inside buildings;
- Scenario B2 — gardeners;
- Scenario B3 — general workers performing intrusive works in the site.

For employees working indoors (Scenario B1) ingestion of soil/dust particles and dermal contact with particles presented potential exposure path-ways. Inhalation of suspended particles was also considered but proved to be of no significance relative to other pathways. For assessment of indoor exposure, it was necessary to relate indoor contaminant concentrations to outdoor concen-trations in soil. According to Calabrese and Stanek[10], this proportion was taken as 30%. Exposure duration and frequency for the oral and dermal pathways were set as 260 days per year (based on the planned shift rota of six working days, two days off and 20 days holidays) and 15 years, respectively (as an upper estimate for the employment duration by the client). According to the US EPA[3] and Ihme[5] the soil ingestion rate was set at 50 mg per working day and a dermal dust cover of 1 mg per cm^2 per day was chosen. The potentially exposed skin surface was 1688 cm^2. This was equivalent to the area of half the head and both hands. The exposure assumptions are listed in Table 15.2.

Another relevant receptor group were gardeners (Scenario B2) who would maintain the landscaped factory surroundings. Based on normal garden-ing activities such as pruning, weeding, digging and grass cutting, the oral and dermal exposure to soil/dust particles were regarded as being most relevant. In-halation of dust particles was not expected to provide a significant exposure route. The exposure times of such, often subcontracted, personnel was assumed to be significantly shorter than for factory workers (four days per month during five years, see Table 15.2, page 176). Weather conditions can prevent exposure to soil because of rainfall, snow cover or frost, so a climatic factor of 0.5 was introduced (that is, it was assumed that exposure could only occur during 50% of the total working time). On the other hand, gardeners often wear less clothes than workers, especially in summer. Therefore additional dermal exposure of lower arms and lower legs was considered possible (summing up to 5318 cm^2 unprotected skin area).

In the last scenario (B3) workers who perform intrusive work in the ground (for example, repairing or installing buried services) were considered. For such work performed during good weather conditions, no climate-related reduction of the total exposure time was made and the time fraction when exposure is possible due to climate was set at 100%. Besides oral and dermal contact with soil, inhalation of suspended particles was regarded to be likely. This assumption was based on a scenario in which workers excavate or work within a trench. Dust concentration was calculated according to Hawley[11] by a transfer factor of 2.4×10^{-7} kg/m^3 which describes the contaminant partitioning between soil and air in a dusty environment. Exposure time was taken as 20 working days during two months per year for five years. As a worst case, it was assumed that the skin area of the head, neck, hands, arms, lower legs and trunk (12,631 cm^2) was exposed. All exposure parameters are listed in Table 15.2.

Together with the toxicological criteria (Table 15.3), health protective soil values were calculated. In Scenario B1 soil values for all compounds are based on oral and dermal exposure. The same is true for Scenario B2 and in Scenario B3 inhalation of suspended particles was also included. For arsenic, being a carcinogen, exposure in all three scenarios had to be averaged over a lifetime and evaluated by the potency factor. As there is no potency factor for the dermal pathway, the oral potency factor may serve as a surrogate value, according to the US EPA.

The other metalloids are characterized by their non-carcinogenic effects. This implies that the reference dose had to be applied for toxicological evaluation with exposure averaged over the actual exposure duration. For lead a different toxicological criterion was applied (the provisional tolerable weekly intake (PTWI)) — there is no RfD value available from the US EPA. The carcinogenic effect of cadmium via inhalation proved to be negligible compared with its non-carcinogenic effects.

For mineral oil the toxicological criterion for n-hexane was applied. This provides a conservative assessment of the non-carcinogenic properties of the whole product (PAH and benzene, both proven carcinogens, were evaluated

179

separately). PAHs were evaluated by using benzo(a)pyrene (BaP) as an indicator substance. Therefore the soil values in Scenarios B1 to B3 apply to the BaP concentration but ensure that there is no health risk from the total group of PAHs. This evaluation assumes a typical PAH profile that causes a tenfold higher oral risk by all PAHs compared to BaP alone and a 50-fold higher inhalative risk, respectively.

As the most sensitive receptor must be protected, the lowest soil value is selected from the three scenarios to drive the decision on a potential remedial goal. Arsenic, for example, provided the lowest soil value in Scenario B1. There the maximum tolerable soil concentration proves to be 90 mg/kg. To give another example, mineral oil should not exceed a concentration of 15,000 mg/kg (assuming the oil does not contain a significant fraction of product with short chain length and low boiling point and therefore high volatility). Which scenario proves to be the most stringent, and therefore critical, depends on various factors such as the toxicological effect evaluated (carcinogenic/non-carcinogenic), the pathways considered and the mutual proportions of toxicity data on the single pathways. The most stringent soil values are shown in bold in Table 15.4.

DISCUSSION

The two case studies have introduced examples of how to assess risk that is related to soil/groundwater contamination (Scenarios A1 and A2). They also show how to derive health-protective soil concentrations (soil values) that do not pose a significant health risk to a potential human receptor and which may serve as remedial goals (Scenarios B1 to B3). Both approaches provide results that enable the site environmental manager to assess release of contaminants to the environment, and gain useful information for the environmental management process. Due to their site- and use-specificity these results are much more appropriate than generic criteria to evaluate the risk contaminated land poses to humans and/or the environment.

In the past the UK ICRCL values as well as the Dutch criteria were often applied in order to provide some guidance. Neither of them has any force in law in the UK. Both lists have specific drawbacks. While the ICRCL values provide a few use-specific differentiations, there are none within the Dutch List which instead requires a multifunctional end use. This directly contrasts with the UK Government's 'suitable to use' approach. In Table 15.5 (see page 182)

TABLE 15.4
Health protective soil concentrations for an industrial site

Compound	Soil value (remedial goal), mg/kg		
	Scenario B1*	Scenario B2**	Scenario B3***
Arsenic	**90**	410	110
Cadmium	130	270	**90**
Copper	9300	19,600	**6900**
Lead	**8400**	53,100	18,400
Zinc	75,600	159,200	**55,900**
Mineral oil[1]	46,200	56,500	**15,000**
PAHs[2]	**1.8**	8.0	2.2

Bold type indicates the most stringent soil values
* Scenario B1 = workers — indoor exposure
** Scenario B2 = gardeners
*** Scenario B3 = workers performing intrusive work in the ground
[1] n-hexane is taken as a surrogate compound for mineral oil (excluding BTEX and PAHs); soil value refers to a non-volatile species of mineral oil product
[2] the soil value refers to benzo(a)pyrene as an indicator substance but evaluates the risk from all PAHs

site-specific concentrations, health-protective soil values, ICRCL and Dutch criteria are presented together. The risk-based soil values are higher than the Dutch and intervention values as well as the ICRCL action values. Except for PAHs there are no threshold values given by the ICRCL. It has to be noted that the data for PAHs are of limited comparability. The soil value refers to the BaP concentration alone, while the sum of all PAHs may be five to ten times higher.

Another problem with generic criteria is the limited scope of all criteria lists. This problem can be overcome by the risk assessment process which is able to evaluate any compound for which at least some basic toxicological knowledge is available.

It should be considered that the soil values introduced in this chapter can serve as remediation goals. However, sometimes there may be reasons to

TABLE 15.5
Concentrations on site, lowest remediation goals opposed to ICRCL values and Dutch criteria

Compound	Soil concentration[1], mg/kg Average	Max.	Lowest soil value[2]	ICRCL values, mg/kg Threshold	Action	Dutch current criteria[3], mg/kg G-value	I-value
Arsenic	10	25	90	40[4]	–	42	55
Cadmium	1	10	90	15[4]	–	6.4	12
Copper	140	3300	6900	130[5]	–	113	190
Lead	250	3600	8400	2000[4]	–	308	530
Zinc	500	9600	55,900	300[5]	–	430	720
Mineral oil[6]	1200	2100	15,000	–	–	2525	5000
BaP	2	25	1.8[7]	–	–	–	–
PAHs	–	–	–	1000[8]	10,000[8]	21	40

BaP = benzo(a)pyrene
ICRCL = Interdepartmental Committee on the Redevelopment of Contaminated Land
[1] all values have been rounded to the nearest whole number
[2] as the most sensitive receptor has to be protected, the lowest value has to be taken from Table 15.4
[3] Dutch criteria — that is, the guide value (G-value) and intervention value (I-value) refer to a standard soil type
[4] applies to parks, playing fields, open space
[5] applies to any uses where plants are to be grown
[6] soil value refers to a non-volatile species of mineral oil product
[7] refers to benzo(a)pyrene as an indicator but applies to the risk from all PAHs
[8] applies to landscaped areas, buildings, hard cover

lower the values due to further objectives. It may be necessary — especially for recreational and residential sites — to include ecotoxicological considerations. Other aspects, such as aesthetic effects (for example, odour threshold) may also need to be considered.

Finally the risk assessment process is an 'open book' and therefore defendable. This also means that details (assumptions, parameters) can be discussed between parties (for example, the environmental health manager and the local authority) and evaluated with respect to their effect on the risk finally estimated. An agreement on the acceptable risk has to be found which in most cases reflects the current recommendations of such authorities as the UK Department of the Environment, the World Health Organization, the US EPA and other bodies. The risk assessment approach is subject to continued refinement and improvement from the increasing amount of research into the subject.

REFERENCES IN CHAPTER 15

1. DoE (UK Department of the Environment), 24 November 1994, Framework on contaminated land, *News Release 654*.
2. NAS (National Academy of Sciences), 1983, *Risk Assessment in the Federal Government: Managing the Process* (National Academy of Sciences, National Academy Press, Washington DC, USA).
3. US EPA (US Environmental Protection Agency), 1989, Risk assessment guidance for Superfund, Volume I, *Human Health Evaluation Manual (Part A), EPA/540/1-89/002* (US EPA, Washington DC, USA).
4. McKone, T.E., 1988, Methods for estimating multi-pathway exposures to environmental contaminants, *Final Report, Phase II, UCRL–21064* (Lawrence Livermore National Laboratory, USA).
5. Ihme, W., 1994, *General Model for Quantification of Exposure of Humans to Contaminated Soils* (in German) (Shaker Verlag, Aachen, Germany).
6. Smucker, S.J., 1994, *Region IX, Preliminary Remediation Goals (PRGs), First Half 1994* (US EPA, Region IX, San Francisco, USA).
7. WHO (World Health Organization), 1972, Evaluation of certain food additives and contaminants, *16th Report of the Joint FAO/WHO Expert Committee on Food Additives, Technical Report Series No. 505* (World Heath Organization, Geneva, Switzerland).
8. Slooff, W., Janus, J.A., Matthijssen, A.J.C.M., Montizaan, G.K. and Ros, J.P.M. (eds), 1989, Integrated criteria document PAHs, *Report No. 758474011* (National Institute of Public Health and Environmental Protection (RIVM), Bilthoven, The Netherlands).
9. Grimmer, G., Brune, H., Dettbarn, G., Jacob, J., Misfeld, J., Mohr, U., Naujack, K.W., Timm, J. and Wenzel-Hartung, R., 1990, Contribution of polycyclic aromatic hydrocarbons and other polycyclic aromatic compounds to the carcinogenicity of combustion source and air pollution, in Waters, M.D. (ed), *Genetic Toxicology of Complex Mixtures* (Plenum Press, New York, USA), 127–140.

10. Calabrese, E.J. and Stanek, E.J., 1992, What proportion of household dust is derived from outdoor soil?, *Journal of Soil Contamination*, 1 (3): 253–263.
11. Hawley, J.K., 1985, Assessment of health risk from exposure to contaminated soil, *Risk Analysis*, 5 (4): 289–302.

16. INTEGRATED POLLUTION CONTROL — APPLICATION OF PRINCIPLES TO ESTABLISH BPEO AND BATNEEC

Stefan Carlyle

This chapter describes a new procedure being introduced by Her Majesty's Inspectorate of Pollution (HMIP) to carry out environmental, economic and best practicable environmental option (BPEO) risk assessments for integrated pollution control (IPC).

INTEGRATED POLLUTION CONTROL

The statutory basis for IPC is provided in Part 1 of the Environmental Protection Act 1990 (EPA90). IPC requires that no prescribed process can be operated without a prior authorization from HMIP. In setting the conditions to be attached to an authorization, Section 7 of the Act places HMIP under a duty to ensure that certain objectives are met. The conditions should ensure that:

• the best available techniques (both technology and operating practices) not entailing excessive cost (BATNEEC) are used to prevent or, if that is not practicable, to minimize the release of prescribed substances into the medium for which they are prescribed; and to render harmless both any prescribed substances which are released and any other substances which might cause harm;

• releases do not cause, or contribute to, the breach of any direction given by the Secretary of State to implement European Community or international obligations relating to environmental protection, or any statutory environmental quality standards or objectives, or other statutory limits or requirements;

• when a process is likely to involve releases into more than one environmental medium (which will probably be the case in many processes prescribed for IPC), the BPEO is achieved — that is, the releases from the process are controlled through the use of BATNEEC to give the least overall effect on the environment as a whole.

IPC APPLICATIONS

In applying for an IPC authorization an operator should:

• provide full information on the selection of primary process, particularly for a new plant;

- provide evidence that the requirement to use BATNEEC will be met;
- select a combination of primary process, pollution abatement techniques and waste treatment and disposal which constitutes the BPEO;
- provide a justification for the BPEO selected and for all likely releases.

ENVIRONMENTAL ASSESSMENTS FOR IPC

The Centre for Integrated Environmental Risk Assessment (CIERA) of HMIP is developing principles and practical procedures for the assessment of environmental harm arising from releases from prescribed processes and to determine the BPEO and the site-specific BATNEEC option for a particular IPC process.

Such a procedure might consist of a number of elements. For example:

- the identification and quantification of releases;
- determining whether releases comply with statutory emission limits;
- a scoping exercise to identify environmentally significant releases;
- determining whether the releases comply with statutory environmental quality objectives;
- determining the environmental acceptability of the releases;
- identifying the BPEO from a number of environmentally acceptable process/abatement options;
- identifying the process and environmental monitoring requirements.

PRINCIPLES OF THE IPC ENVIRONMENTAL ASSESSMENT PROCEDURE

DEFINITION OF HARM

It is clear from the objectives given in Section 7 of EPA90 that a means for assessing harm is required by HMIP for use in making regulatory judgements. Within the context of the Act, 'harm' means:

'Harm to the health of living organisms or other interference with the ecological systems of which they form a part and, in the case of man, includes offence caused to any of his senses or harm to his property; and 'harmless' has a corresponding meaning.'

However, the Act does not define the nature of the effects which may be considered harmful or the level in the environment at which they may occur. Therefore, to overcome these difficulties a practical approach to the assessment of harm has been devised and is described in the following section.

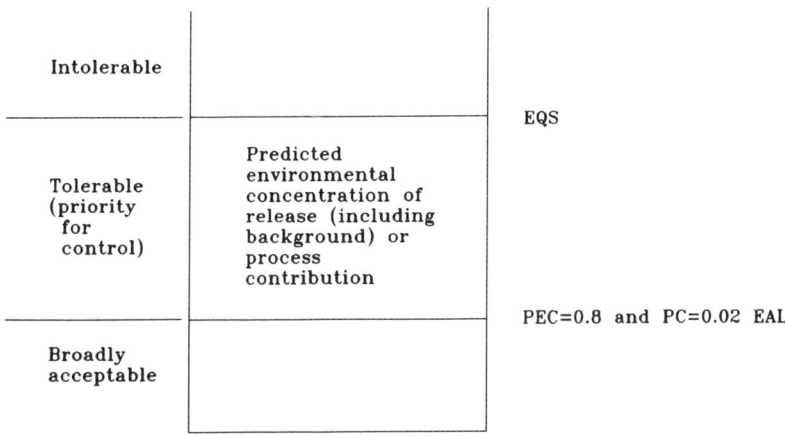

Figure 16.1   Principles for assessing the level of harm caused by individual releases.

The principles for assessing the level of harm caused by individual releases is illustrated in Figure 16.1. The intolerable level is defined by either a
statutory limit such as an environmental quality standard (EQS) for releases to
air or an environmental quality objective for those to water. In the absence of a
statutory limit for a substance, an interim environmental assessment level
(EAL) will be set by HMIP to enable comparisons to be made. An EAL is the
concentration of a substance which HMIP regards as a comparator value to enable a comparison to be made between the environmental effects of different
substances within and across environmental media. Where an EQS is available
it is normally used as an EAL.

HMIP Guide Levels can be used in assigning priorities for control; for
example, if a predicted maximum ground level concentration is high compared
with the EAL (more than 80%) the release might command priority in generating alternative process options. Similarly, if the contribution of an individual
plant to the ambient level of a substance is more than 2% then it might be assigned a higher priority for control.

The threshold value used to determine whether or not a release is significant is set at a level where the effect on the environment can be confidently
considered as negligible. Again a threshold value is being developed by HMIP

as a guide to operators and site inspectors on the significance of releases where detailed environmental assessment is not required.

DEFINITION OF THE BPEO

Where a process involves the release of substances to more than one environmental medium, the Inspector needs to determine whether the proposed operation represents the BPEO for the pollutants concerned. Since EPA90 does not define the BPEO, however, it is proposed to adopt (with minor modification) the definition provided by the Royal Commission on Environmental Pollution in its *12th Report*[1] as follows:

'The BPEO can be considered as the option which, for a given set of objectives, provides the most benefit or least damage to the environment as a whole, at acceptable cost, in the long term as well as the short term, *as a result of releases of substances from a prescribed process*.'

The modification, highlighted by italics, has been added to bring the definition explicitly within the context of Section 7(10) of EPA90, which limits the scope of the BPEO to consideration of substances released by the process.

The approach outlined above to assess the tolerability of releases can be used to assess the BPEO. Where tolerable significant releases occur to more than one environmental medium (see Figure 16.1), process/abatement options can be compared in an overall environmental (BPEO) context. Each significant release can be normalized by expressing the predicted environmental concentration as a percentage (or proportion) of the environmental assessment level. The proportion of the EAL utilized by each significant release can be summed to create a environmental quotient (EQ) for each medium, which is then used to derive an integrated environmental index (IEI) for a specific process/abatement option. This index could be used as the main basis on which to establish the best option from an environmental protection viewpoint. Additional environmental factors would need to be taken into account in ranking alternative process options (see below).

SITE-SPECIFIC BATNEEC

The BPEO ranking derived from the above assessment is combined with the economic or NEEC aspects to determine the site-specific BATNEEC option. A prescribed process should broadly meet the environmental performance of processes identified in Chief Inspector's Guidance Notes (CIGNs) — either those actually described in the relevant note, or one which achieves an equivalent or better level of environmental protection. An operator should indicate the range of process/abatement options considered in deciding on the eventual choice. If an option is proposed that clearly falls below the best environmental option, then an operator would need to justify the choice in terms of cost effectiveness — that is, by comparing the cost and environmental implications of the preferred and any discounted options. It is for HMIP's site inspector to assess whether or not sufficient justification has been provided, perhaps by carrying out a separate assessment.

ENVIRONMENTAL ASSESSMENT PROCEDURE

Having established the main principles, it is necessary to develop a robust regulatory procedure. This should minimize the effort required by operators in preparing an application and by inspectors in assessing and confirming BAT-NEEC, as well as providing an audit trail documenting how the key decisions are reached. The stages in the procedure are as follows:

STAGE I: PRELIMINARY ENVIRONMENTAL ASSESSMENT

Step 1: Identify pollutants released
For a specific (base case) process/abatement option, identify the unavoidable releases of prescribed (and other potentially harmful) substances. Establish the release rates under start-up, normal and abnormal operation.

Step 2: Significance test
Carry out a significance test to eliminate trivial releases from further consideration in the assessment. This will be based on advice provided by HMIP on levels (mass and/or concentration) of substances in releases which might be regarded as insignificant.

Step 3: Comparison with Chief Inspector's Guidance Notes
CIGNs provide an indication of best practice for particular prescribed processes;

the process in question should achieve the levels of environmental protection advocated in the relevant CIGN.

Step 4: Chimney height determination
Chimney height determination ensures that releases are rendered harmless so far as short-term maximum ground level concentrations are concerned.

Step 5: Compute predicted environmental concentrations (PECs)
Use basic dispersion models to predict the maximum environmental concentration of each release. Add to this the actual or estimated background concentration of the pollutant at that point.

Compare the predicted environmental concentration for each significant release with any statutory limit (EQS), or where this is absent with the relevant environmental assessment level (EAL). (Those options resulting in a breach of the statutory limit are automatically discounted). Where a release results in a particularly high PEC compared with the EQS or EAL (typically greater than 80%), the Inspectorate might wish to examine this in more detail to reduce any uncertainties and to examine alternative process/abatement options.

STAGE 2: PROCESS OPTION GENERATION
The information gathered in Stage 1 about the environmental risks posed by the basic process can be used, together with other relevant factors, to generate alternative process options. Based on Stage 1, relevant environmental factors might include:
• reducing releases of substances which might lead to a breach in an EQS;
• reducing releases where the predicted environmental concentration is high compared to the EAL — for example, more than 80% of an EAL;
• reducing releases where the process contribution would be more than 2% of an EAL;
• reducing releases to ensure that sensitive ecosystems are not harmed.

Other factors might also be relevant to generating options for BPEO assessment, including:
• health and safety;
• physical space;
• loss of amenity;
• energy consumption.

190

STAGE 3: BPEO ASSESSMENT

Step 1: For significant releases establish the normalized environmental quotient and calculate the integrated environmental index (IEI)
The PECs for significant release should be confirmed using more comprehensive predictive models. Releases are normalized as a percentage of the statutory limit (or interim EAL) to give the environmental quotient (EQ). EQs for significant releases to all media are summed to give the integrated environmental index. The derivation of an IEI can be summarized as follows:
(i) For a tolerable single release:

$$EQ = \frac{PEC}{EAL}$$

(ii) EQ for all tolerable significant releases to an environmental medium:

$$EQ^{(air)} = EQ_a + EQ_b \ldots + EQ_i \text{ (for a + b } \ldots \text{ i releases)}$$

(iii) Integrated environmental index:

$$IEI_{(index)} = EQ^{(air)} + EQ^{(water)} + EQ^{(land)}$$

(iv) Best environmental option (BEO) from a range of environmentally tolerable options:

$$BEO = \text{lowest integrated environmental index}$$

Step 2: Assessment of other environmental factors
Other relevant factors should be assessed to check the validity of the ranking of options based on the IEI. These might include short-term effects, global warming effects, potential to generate ozone, waste arisings and so on. These would need to be taken into account by carrying out a kind of multi-attribute analysis to rank the options in terms of environmental effects.

SHORT-TERM EFFECTS
Different process options may lead to variations in the pattern of releases. For example, a process operated intermittently may give lower annual concentrations compared to one run continuously, but an increased frequency of short-term peaks may be the result. The assessment of short-term releases should therefore

be an integral part of the environmental assessment under IPC for both new and existing processes.

ASSESSMENT OF GLOBAL WARMING POTENTIAL (GWP) OF RELEASES

The release of carbon dioxide, water vapour, chlorofluorocarbons (CFCs), methane and nitrous oxide may contribute to global warming. In addition, tropospheric ozone (that is, ozone in the bottom 8–16 km of the atmosphere) also acts as a greenhouse gas, but the magnitude of this effect is still under investigation. It is important that the release of these gases is minimized wherever possible. Due to the nature of the effects arising from these pollutants, it is not possible to incorporate them directly in the environmental index. Instead, process options should be ranked according to their potential to contribute to radiative forcing (global warming) expressed in carbon dioxide equivalents.

ASSESSMENT OF THE POTENTIAL FOR OZONE GENERATION

Ozone is a highly reactive pollutant which may exert a number of damaging effects on human health, vegetation and materials. The production of ozone in the troposphere involves the action of sunlight on hydrocarbons, usually referred to as volatile organic compounds (VOCs), and oxides of nitrogen (NO_x). The availability of NO_x downwind of a source controls the spatial extent of the area within which raised ozone concentrations may be generated.

There is a large variation between the importance of individual VOCs in their potential for ozone generation, depending on their reactivity with hydroxyl (OH) radicals and the subsequent production of peroxy (RO_2) radicals. In order to assess the relative effect of different hydrocarbons in the episodic production of ozone, and provide a basis for their control, the UNECE VOC convention[2] has proposed the concept of the photochemical ozone creation potential (POCP). The POCP is defined as the change in photochemical ozone production due to a change in emission of that particular VOC. The POCP may be determined by photochemical model calculations or by laboratory experiments.

Estimated individual POCP values vary both temporally and spatially depending on the VOC composition of the modelled air parcel, the assumed meteorological conditions and NO_x concentrations. However, although there is considerable uncertainty over individual POCP values, the approach can be used to classify VOC species according to their importance in ozone production and average values assigned to each class of compound. It can be used, for example, to rank process options according to the overall POCP of the option.

ASSESSMENT OF WASTE ARISINGS

Many processes will generate quantities of solid or liquid waste — that is, material which is not released to air or water. These wastes may be treated or disposed of on site or removed from the plant for treatment or disposal elsewhere.

Incorporating such waste arisings into the calculation of an IEI is less straightforward than is the case with releases to air or water. This is because their environmental effects are likely to depend on a range of more complex interactions. A separate means of comparing this aspect of different process options is required, and this can be achieved by assessing each waste arising on the basis of quantity and relative hazard potential. The relative hazard potential is determined by physical, chemical and biological characteristics of the waste.

A number of hazard assessment schemes can be found in the literature. The one proposed in this document is based on that developed by the UK Government Industry Working Group on Priority Setting and Risk Assessment[3].

The scheme is based on a number of parameters:

• toxicity (to mammals and aquatic organisms);
• potential for bioaccumulation;
• degradation (in soil/water);
• other physical characteristics such as solubility, adsorption potential and volatility.

Each of these parameters is scored and a total score is obtained by multiplying the individual scores. The score obtained represents the potential hazard for a unit quantity of the substance concerned. The 'unit hazard' can then be weighed by the quantity generated to obtain a final score for the substance concerned. The final scores for all the components of a waste are summed to give an overall hazard score for the process being considered. Different process options producing different wastes can be ranked according to their overall hazard scores.

OTHER ENVIRONMENTAL FACTORS

The above factors are neither exhaustive nor exclusive. Other factors that may be relevant at a particular site include odours, visible plumes from chimneys, releases of dioxins or furans, releases of acid gases where critical loads might be exceeded, non-routine releases and so on. But not all of these factors will be relevant to any one site and judgement must be used in deciding which factors are relevant, both in preparing and evaluating an IPC application.

STAGE III: DETERMINE SITE-SPECIFIC BATNEEC OPTION
From the range of environmentally tolerable options generated under Stages 1 and 2, the next step is to select the site-specific BATNEEC option. If the choice has the lowest BPEO index (greatest environmental protection) then no comparative economic analysis is needed. If not, then the reason for rejecting the more environmentally beneficial options should be justified in terms of cost effectiveness.

No formal rule for the choice of BATNEEC can be prescribed. (An operating company must reach its own view as to which of the potential process/abatement options is the site-specific BATNEEC). However, the application of cost-effectiveness analysis assists the operator by reducing the number of options. This can be achieved in two ways:

• options that are cost-effective can be identified. A cost-effective BATNEEC process/abatement option is defined as the one which achieves a given level of pollution control at least cost;

• the costs of achieving more stringent levels of pollution control can be shown. This information illustrates any significant 'break point' beyond which reductions in pollution can only be achieved at greater incremental cost.

When identifying the least cost option, it is important to note that the cost-effectiveness result may be sensitive to the context in which an individual abatement technique is applied. For example, plant is designed for a specified range of operating conditions, and to operate beyond that range or under different conditions may entail additional operating and/or capital costs.

In identifying a 'break point' where the costs of further reductions in pollution potential start to rise significantly, it is important to have a quantitative measure of the environmental consequences of the releases involved. Where the choice of the preferred option has been based largely on the long term consequences of the releases as indicated by the environmental index, then this may provide a convenient measure of the overall pollution potential of a process. When the preferred option has been selected on the basis of other environmental factors, it may be more appropriate to use different measures of pollution potential — such as the quantity of individual substances removed or the global warming potential of different abatement options.

Two main techniques can be used to identify cost-effective treatment or abatement options:

• annualized costs of different levels of pollution control can be plotted on a diagram, as illustrated in Figure 16.2; or

- incremental cost-effectiveness can be presented arithmetically to compare the costs and abatement of a process option with those of the next more environmentally harmful option. This can be expressed more formally as:

Incremental cost =

$$\frac{\text{Cost option 1} - \text{Cost option 2}}{\text{Difference in pollution potential (Option 2} - \text{Option 1)}}$$

Figure 16.2 shows the costs and environmental effects for a range of different process options/techniques for a hypothetical product. A BPEO or environmental index (0– 4) derived from emission levels is shown for each option, including the current operation which is the base case. The operator might consider option 3 as representing the site-specific BATNEEC even though option 4 has a better BPEO index value, on the basis of the higher incremental cost entailed at that site in attaining the environmental benefit of option 4.

Table 16.1 (see page 196) is an example of how the incremental costs of pollution control can be demonstrated. Other things being equal, in this case it might be concluded that the site-specific BATNEEC is option 3, because of the doubling of incremental cost to achieve any greater improvement.

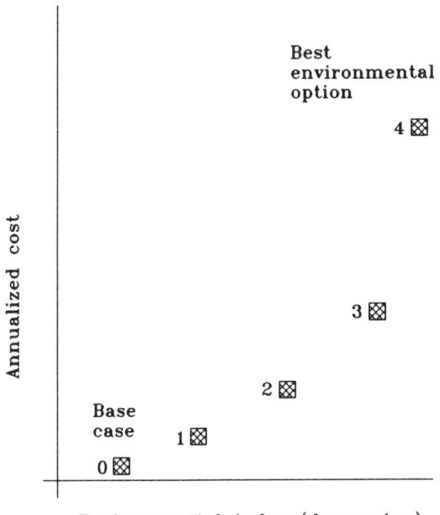

Figure 16.2 Principles for establishing the site-specific BATNEEC option.

TABLE 16.1
Illustration of the incremental cost of pollution control

Option	Equivalent annual cost	Incremental improvement in IEI	Incremental cost
Base case (uncontrolled)	£0	–	–
Option 1	£20,000	1	£20,000
Option 2	£50,000	1	£30,000
Option 3	£100,000	1	£50,000
Option 4	£200,000	1	£100,000

CONCLUSIONS

HMIP's work on developing a straightforward environmental assessment framework for IPC should help both industry and the Inspectorate in knowing exactly what information should be provided. A consultation document that expands on the assessment principles has been issued[4]. The comments received from some 100 organizations, together with the results of comprehensive case studies, have been taken into account in finalizing the technical guidance note. HMIP will publish this shortly.

REFERENCES IN CHAPTER 16
1. Royal Commission on Environmental Pollution, 1988, *12th Report* (HMSO).
2. United Nations Economic Commission for Europe (UNECE), 1991, *Protocol to the 1979 Convention On Long-Range Transboundary Air Pollution Concerning the Control of Emissions of Organic Compounds or their Transboundary Fluxes* (UNECE, Geneva).
3. UK Government Industry Working Group on Priority Setting and Risk Assessment, 1991 (DoE).
4. HMIP, April 1994, *Environmental, Economic and BPEO Assessment Principles for Integrated Pollution Control — A Consultation Document.*

FURTHER READING

The Environmental Protection (Prescribed Processes and Substances) Regulations 1991 SI No 472, as amended by the Environmental Protection (Prescribed Processes and Substances) (Amendment) Regulations 1992, SI No 614 (HMSO).

DoE, 1991, *Integrated Pollution Control — A Practical Guide.*

LIST OF ACRONYMS

ADI — acceptable daily intake

BATNEEC — best available techniques not entailing excessive cost
BPEO — best practicable environmental option
BS — British Standard
BSI — British Standards Institution

CEN — Comité Européen de Normalisation
CIGN — Chief Inspector's Guidance Note
CMP — catchment management plan
COSHH — Control of Substances Hazardous to Health Regulations

DoE — UK Department of the Environment

EA — environmental auditing
EAL — environmental assessment level
EMA — Eco-Management and Audit
EMAS — Eco-Management and Audit Scheme
EMS — environmental management system
EPA90 — UK Environmental Protection Act 1990
EPE — environmental performance evaluation
EQ — environmental quotient
EQS — environmental quality standard
EVABAT — economically viable application of best available technology

GLC — ground level concentration

HMIP — Her Majesty's Inspectorate of Pollution
HSW74 — Health and Safety at Work Act 1974

IEI — integrated environmental index
IPC — integrated pollution control
ISO — International Organization for Standardization

LULU — locally unacceptable land use

MAFF — UK Ministry of Agriculture, Fisheries and Food

NACCB — UK National Accreditation Council for Certification Bodies
NRA — UK National Rivers Authority

OES — occupational exposure standard

PEC — predicted environmental concentration
POCP — photochemical ozone creation potential

QA — quality assurance

RfD — reference dose

SWQO — statutory water quality objectives

TQM — total quality management

WQO — water quality objectives
WRA — Waste Regulation Authorities